ESSAYS IN THE ECONOMICS OF INVASIVE SPECIES MANAGEMENT

AMITRAJEET A. BATABYAL
ROCHESTER INSTITUTE OF TECHNOLOGY, USA

NorthAmerican
Business Press
Atlanta – Seattle – South Florida – Toronto

I am monarch of all I survey;
My right there is none to dispute;
From the centre all around to the sea
I am lord of the fowl and the brute.
O Solitude! Where are the charms
That sages have seen in they face?
Better dwell in the midst of alarms,
Than reign in this horrible place.

—*The Solitude of Alexander Selkirk*, William Cowper

North American Business Press, Inc
Atlanta, Georgia
Seattle, Washington
South Florida
Toronto, Canada

Essays in the Economics of Invasive Species Management
ISBN: 9780988919310

Along with trade books for various business disciplines, the North American Business Press also publishes a variety of academic-peer reviewed journals.

Library of Congress Control Number: 2013932945
Library of Congress
Cataloging in Publication Division
101 Independence Ave., SE
Washington, DC 20540-4320
Printed in the United States of America

First Edition

CONTENTS

ACKNOWLEDGMENTS

Of the fifteen chapters in this book, thirteen have appeared previously in different journals. Therefore, I would like to take this opportunity to express my appreciation to various publishers for granting me permission to reprint these previously published chapters. First, Chapter 2 "International Trade and Biological Invasions: A Queuing Theoretic Analysis of the Prevention Problem" originally appeared in the *European Journal of Operational Research*, Vol. 170, pp. 758-770, 2006, Chapter 5 "Trade, the Damage from Alien Species, and the Effects of Protectionism Under Alternate Market Structures" originally appeared in the *Journal of Economic Behavior and Organization*, Vol. 70, pp. 389-401, 2009, Chapter 8 "Some Statistical Properties of a Generic Container Inspection Policy in Invasive Species Management" originally appeared in *Ecological Economics*, Vol. 60, pp. 1-4, 2006, Chapter 9 "Some Probabilistic Attributes of Inspection Policies in Alien Species Management" originally appeared in *Ecological Economics*, Vol. 60, pp. 5-8, 2006, and I thank Elsevier for their kind permission to reprint these four papers. Second, Chapter 6 "Maritime Trade, Biological Invasions, and the Properties of Alternate Inspection Regimes" originally appeared in *Stochastic Environmental Research and Risk Assessment*, Vol. 19, pp. 184-190, 2005, Chapter 7 "On Container Versus Time Based Inspection Policies in Invasive Species Management" originally appeared in *Stochastic Environmental Research and Risk Assessment*, Vol. 19, pp. 340-347, 2005, Chapter 10 "The Infinitesimal, the Deterministic, and the Probabilistic: Alternate Container Inspection Policies in Invasive Species Management" originally appeared in *Biological Invasions*, Vol. 8, pp. 1663-1671, 2006, Chapter 11 "An Analysis of Economic Cost Minimization and Biological Invasion Damage Control Using the *AWQ* Criterion" originally appeared in the *Annals of Regional Science*, Vol.

41, pp. 639-655, 2007, Chapter 14 "Optimal Resource Management in the Presence of a Deleterious Alien Species: A Stochastic Model for an Orchard" originally appeared in *Letters in Spatial and Resource Sciences*, Vol. 1, pp. 107-115, 2008, Chapter 15 "An Approach to the Management of Orchards That Are Vulnerable to Attack by Invasive Species" originally appeared in *Letters in Spatial and Resource Sciences*, Vol. 2, pp. 123-131, 2009, and I thank Springer for granting me permission to reprint these six papers. Third, Chapter 3 "A Rationale for the Differential Regulatory Treatment of Imports When Invasive Species Are a Potential Problem" originally appeared in *Studies in Regional Science* and I thank the Japan Section of the Regional Science Association International for allowing me to reprint this paper. Fourth, Chapter 4 "A Dynamic Bertrand Game Model of Trade, Threats, and Export Certification in Alien Species Management" originally appeared in the *Journal of Economic Research*, Vol. 12, pp. 65-78, 2007, and I thank the Hanyang Economic Research Institute for permitting me to reprint this paper. Finally, I note that Chapter 12 "A Theoretical Analysis of Random Inspections and Fines in Invasive Species Management" initially appeared in *Economics Bulletin*, Vol. 17, pp. 1-9, 2008.

This book would not have seen the light of day without the assistance of two groups of individuals. The first group consists of my valued coauthors Hamid Beladi, Gregory J. DeAngelo, Won W. Koo, Seshavadhani Kumar, Dug Man Lee, Christopher D. Lominac, Peter Nijkamp, and Seung Jick Yoo, and I thank them all for allowing me to include our jointly authored papers in this book. The second group consists of my indefatigable Assistant Cassandra Shellman and I thank her for all manner of assistance with this book. In addition, I acknowledge financial assistance from the Gosnell endowment at RIT.

My wife Swapna B. Batabyal and my daughter Sanjana S. Batabyal keep me company, keep me sane, and lose no opportunity to remind me that there is more to life than reading and writing about economics. I thank them both for their love and support. Finally, I dedicate this book to the memory of my late brother Balarka Amarnath Batabyal.

CHAPTER 1

INTRODUCTION TO ESSAYS IN THE ECONOMICS OF INVASIVE SPECIES MANAGEMENT

Abstract. We begin by outlining some of the more salient issues in contemporary research on the management of invasive species. Next, we note that even though there is a sizeable literature in both biology and economics on invasive species management, until very recently, there has been a significant dearth of *theoretical* research in this subject. Given this state of affairs, a key objective of this book is to show how dynamic and stochastic approaches from economics and operations research can be fruitfully used to first model and then analyze a whole host of invasive species related problems that also have noteworthy public policy ramifications. This book is concerned with the *economics* of invasive species management. To this end, we focus attention on the three topics of international trade, inspections, and agriculture. Following this introductory chapter, the next four chapters study the many and varied nexuses between international trade and invasive species related problems. The next eight chapters focus on inspections, an important but nonetheless insufficiently studied topic in invasive species management. Finally, the last two chapters dwell on invasive species related problems in the context of agriculture.

1. Introduction

There is no doubt that the phenomenon we now call globalization has brought many benefits to the citizens of our heterogeneous world. *Inter alia*, this phenomenon has resulted in the increased mobility of humans, goods, and services from one part of the world to another. International trade in both final goods and services and in various factors of production has risen, and this rise has generally been welfare improving for all the involved parties. This rise in international trade in particular and of mobility more generally has also led to the spread of all manner of animals, plants, and other organisms from places where they are native to new locales, where they are not. As noted by Vitousek *et al.* (1997), Mooney (1999), and others, it is salient to comprehend that although this movement and spread of animal and plant species from one part of the world to another predates the modern era of globalization, the phenomenon of globalization has accelerated this mobility.

Animal and plant species that have moved from a part of the world where they are native to another part of the world where they are not are said to be *invasive* species. Such species are also referred to as "alien," as "exotic," and as "non-native" species. For the purpose of consistency, we shall use the term "invasive," in this chapter. Olson (2006) points out that this movement of animal and plant species can occur naturally, as a result of specific ecological processes or as a result of deliberate human activity. In addition, even if we restrict attention to invasive species introductions that are the result of conscious human activity, such introductions can be either intentional or unintentional in nature.[1]

Humans have been intentionally introducing animal and plant species into parts of the world where they are not native, for many decades. These intentional introductions broadly fall into two categories. The first category consists of introductions that are largely ornamental in nature. Exotic pets such as fish, parrots, and all manner of orchids readily come to mind.[2] The second category is made up of introductions that form the basis of what is now biological control. The basic objective here is to use a specific animal or plant species that is a predator of another prey species whose population is sought to be controlled. McConnachie *et al.* (2003) review ten benefit-cost studies of biological control programs and note that such programs can be very successful. However, the work of Nogales *et al.* (2004) shows that biological control programs can also go awry and give rise to unintended consequences. Most countries today have protocols for dealing with the intentional introductions of invasive species, although it should not go

1The invasive species themselves can be either terrestrial or aquatic and sometimes it is important to distinguish between these two broad types of invasive species. See Lovell *et al.* (2006) and Olson (2006) for more on this distinction. However, for the purpose of the discussion in this introductory chapter, we shall not be concerned with this distinction.

2See Brown (2006) for a detailed discussion of the many problems caused by exotic pets in the United States.

unsaid that the quality and the effectiveness of these protocols vary considerably.

This book is *not* about the intentional introductions of invasive species. Instead, the focus here is exclusively on *unintentional* introductions and on the problems created by such unintentional introductions of invasive species. The extant literature—see Perrings *et al.* (2000), Pimentel *et al.* (2000), and Pimentel *et al.* (2005)—is clear on the point that it is these sorts of introductions that typically give rise to what we now call "biological invasions" and it is these invasions that can be and have been very costly from an ecological and an economic standpoint to human societies. This unsavory state of affairs provides much of the impetus for systematically studying, and thereby managing, biological invasions.

To fix ideas and without loss of generality, suppose that the country of interest for us is the United States. Then, the task of managing invasive species in the United States is made difficult by the fact that our understanding of four points is either non-existent or imperfect. First, we typically know little about the *rate* at which alternate invasive species "arrive" in the country. Second, of the many invasive species that arrive in the US, we do not know much about *how many* will actually be successful in surviving in their new environment. Third, of the invasive species that arrive and manage to survive in their new environment, we do not know which ones are beneficial to humans and which ones are *not*. Finally, with regard to the invasive species that are malign, we have difficulty *monetizing* the economic costs resulting from the damage that they frequently cause.

Given this state of affairs, it should be clear to the reader that invasive species management is fundamentally about effective decision making under *uncertainty*. Once this central point is understood, it makes sense to ask more detailed questions about the previous paragraph's four points. The first such question that we ask in this book concerns what in the literature is sometimes referred to as the "pathways" or the "vectors" by which invasive animal and plant species arrive in a nation, such as the United States. Here, many researchers such as Roberts (1999), Batabyal (2004), McAusland and Costello (2004), and Olson (2006) have shown that the pathway/vector in contemporary times is very frequently *international trade*. This is why, following this introductory chapter, chapters 2 through 5, comprising the second part of this book, focus on the international trade dimension of invasive species management.

Given that international trade is a common vector by which invasive species have been accidentally introduced into the United States, the second question we ask concerns how we might prevent potential invasive species from getting a toehold in their new environment in the United States. There are many responses to this question, but one key practical response involves conducting *inspections* at various ports of entry into the United States. The Animal and Plant Health Inspection Service (APHIS) which is a part of the United States Department of Agriculture routinely conducted inspections at various United States ports of entry until this APHIS inspection function was transferred to the Department of Homeland Security, in 2003. Now, despite the salience of inspections as an effective invasive species management tool, barring a few recent studies such as the one by McAusland and Costello (2004), inspections have received little *theoretical* attention in the literature. Therefore, chapters 6 through 13, comprising the third part of this book, study the design and the properties of alternate inspection regimes in considerable detail.

When invasive animal and plant species are able to get a toehold in their new environment, they can detrimentally affect many sectors of the economy of a nation such as the United States. However, the work of Horan *et al.* (2002), Acquaye *et al.* (2005), and others confirms that agriculture is one sector that has been disproportionately impacted in a negative way by the presence of invasive species.[3] Therefore, the third and final question that is asked and answered by chapters 14 and 15,

3See the April 2006 issue of the *Agricultural and Resource Economics Review* for a more detailed corroboration of this claim.

comprising the fourth and final part of this book, concerns the nature and the properties of the management function when the objective of a manager is to protect one kind of agro-ecosystem, namely, an *orchard* from attack by one or more invasive species. We now proceed to discuss the contributions of the individual chapters in this book. We begin with Part II, which focuses on international trade.

2. International Trade

2.1. Queuing theory and the prevention problem

The setting studied in chapter 2 consists of a stylized, publicly owned seaport in the coastal region of an arbitrary country. Ships with ballast water arrive at this seaport to load cargo and to then transport this cargo to some other seaport. It is also possible that some ships that come into the seaport in question with ballast water will first unload cargo and then load new cargo for shipment to some other seaport in the world. In either case, the point to note is that the arrival of these ships coincides with the arrival of a whole host of possibly dangerous biological organisms. The arrival process of the ships in the seaport constitutes the arrival process for the queuing models that are employed in this chapter. The seaport under study has I inspectors and hence this seaport will be able to simultaneously inspect a maximum of I arriving ships, at any point in time. If more than I ships arrive at the seaport during a particular time interval, then the ships that are not already being inspected must wait in queue. Finally, there is an upper limit *U* on the maximum number of ships that can be allowed to queue in the seaport. The seaport system consists of ships that are being inspected, ships that are waiting in queue, the *I* inspectors, and the seaport manager.

This chapter—and many others later—uses queuing theory to study the prevention of biological invasions by focusing on two specific inspection regimes. In both regimes, the arrival process of ships is Poisson and the inspection times are exponentially distributed. In the first inspection regime, there are *I* inspectors and the upper limit on the maximum number of ships in the seaport is *U*. In the second inspection regime, the number of inspectors and the upper limit on the maximum number of ships in the seaport coincide and they are both denoted by the positive integer *I*. Using the language of queuing theory, the first inspection regime is a *M/M/I/U* model and the second inspection regime is a *M/M/I/I* model.

This chapter's analysis of the above two queuing models reveals three key differences. First, in the *M/M/I/U* inspection regime, the seaport capacity *U* is exogenous in the long run whereas in the *M/M/I/I* inspection regime, this seaport capacity is endogenous. This difference has implications for the choice of the optimal number of inspectors. Second, the effective arrival rates of ships in the *M/M/I/U* and the *M/M/I/I* regimes are dissimilar, and this dissimilarity means that the expressions denoting the marginal benefit from international trade and the marginal cost of precluding biological invasions will be *different,* and hence the optimal number of inspectors and the optimal ship inspection rate in these two inspection regimes can be expected to be distinct. Finally, the expected number of ships in the *M/M/I/U* inspection regime will generally be greater than the expected number of ships in the *M/M/I/I* inspection regime. This means that the volume of economic activity and hence the likelihood of a biological invasion will be greater in the *M/M/I/U* regime and lesser in *M/M/I/I* the regime.

When inspecting ships in the seaport of the country under study, the chapter 2 analysis does not distinguish between the various arriving ships based on their country of origin. However, there is some evidence to suggest that the maritime trade-related risks from invasive species *depend* on the countries from which goods and services are imported into a nation. Given this state of affairs, what are the conditions under which it makes sense for a seaport manager in a particular country to grant favorable regulatory treatment to the imports emanating from some countries and not to those originating from other countries? This salient question is analyzed in chapter 3 of this book.

2.2. Differential treatment of imports

Chapter 3 studies an environment in which there are two exporting countries (*A* and *B)* and one importing country called Home. The analysis is conducted from the standpoint of the manager of a representative seaport in Home. Home imports goods from *A* and *B* and these two nations transport their goods via ships to the seaport under study in Home. Ships from *A* and *B* carry their cargo and ballast water into the seaport in Home. The Home seaport manager's problem is that the cargo and/or the ballast water in these ships may contain potentially damaging invasive species. Therefore, this seaport manager's job is to ensure that all cargo arriving in ships from *A* and *B* in his seaport and the ballast water in these ships are inspected for the presence of possibly noxious invasive species.

Ships from *A* and *B* arrive at the seaport in Home in accordance with independent, stationary Poisson processes—see Ross (1996, pp. 59-97) for a textbook account—with time invariant rates θ_A and θ_B. When a country *i* ship arrives at the seaport for inspection, D_i, $i=A,B$, the dollar value of this ship's imported cargo per unit time is received and noted by the seaport manager. This seaport manager recognizes that the *risk* of a biological invasion in Home as a result of one or more species from either *A* or *B* invading Home is *different*. Hence, the amount of time taken to inspect A and B ships are denoted by the dissimilar random variables S_A and S_B.

The central question studied in chapter 3 is the following. Given D_A, D_B, S_A, and S_B is it possible to find a criterion that can be used by the seaport manager to grant favorable treatment to the imports from one or the other country? In other words, chapter 3 seeks to derive a *country of origin rule* that tells us which country's ships ought to be *favored* by the seaport manager. Mathematical analysis of the queuing theoretic model in chapter 3 provides the answer to the above question. Specifically, it is shown that in a situation of differential regulatory treatment, if $E[S_A]/D_A < E[S_B]/D_B$ then the seaport manager in Home ought to favor A ships over B ships when inspecting ships for potentially damaging invasive species.

To comprehend the logic of the above result, note that there are two aspects to the problem of managing invasive species. First, there is the trade or economic aspect. From this perspective, laborious inspections that impede the trade related flow of economic activities are undesirable. In contrast, the second aspect of the management problem is the biological aspect. In this regard, note that the arrival of the invasive species *cannot* be separated from the arrival of imports. This calls for stringent inspections to eliminate or substantially reduce the likelihood of invasions by invasive species.

This discussion tells us that the economic and the biological aspects of the regulatory problem pull in *opposite* directions. In addition, we know that the risks from invasive species often vary by country. Therefore, chapter 3 focuses on two economic criteria (D_A, D_B) and two proxies for the risk of invasion (S_A, S_B) to answer the "Which country to favor?" question. The result in question tells us that A ought to be favored over B because the ratio of the risk of invasion proxy to the economic criterion ($E[S_A]/D_A$) for A is lower than the corresponding ratio ($E[S_B]/D_B$) for B.

In addition to implementing the above specified rule, an invasive species manager in Home can also take steps to get exporting firms to take prophylactic actions in their own nations to diminish the likelihood of biological invasions in Home. An examination of these sorts of actions involves the analysis of a game theoretic model in which national actions and reactions can be studied in a meaningful manner. Such an analysis is undertaken in chapter 4 of this book.

2.3. Trade, threats, and export certification

By undertaking measures, such as inspections in a foreign country, an exporting firm in this foreign country can *certify* that the container cargo it is shipping to the home country contains no invasive species. However, because certification is *costly*, the foreign firm has little or no incentive to certify its exports. Therefore, the main question explored in chapter 4 is whether it is possible for an invasive species manager in the home country to *threaten* to levy a duty on the foreign firm's exports and thereby get this firm to certify its exports. The potentially threatening invasive species manager does know the *price* of the export good. However, given the steps a particular exporting firm says it has taken to obtain export certification, this manager also has a pretty good idea about the cost of these steps. As such, on the reasonable assumption that exporting firms are profit maximizers, chapter 4 posits that an export good that has undergone certification will be priced *higher* than the same good without certification.

This *differential pricing* idea is exploited in chapter 4 to ascertain whether it is possible for an invasive species manager in an importing nation to threaten to levy a duty on a foreign firm's exports and thereby get this firm to certify its exports. The game theoretic model in this chapter concentrates on two nations called Home and Foreign that engage in Bertrand competition.[4] The Foreign firm exports and Home imports a particular good. The domestic and the import goods are *imperfect* substitutes. The price of the imported good and the domestic good in Home are p and q respectively.

Now, given the period 1 import price chosen by the exporting firm from Foreign, the Home invasive species manager imposes an *ad valorem* duty in period 2, whenever the first period import price is less than a benchmark price. This benchmark price is the Home invasive species manager's best estimate of what the price of the Foreign firm's export good would be if its exports had, in fact, been certified. The invasive species manager does not impose the above *ad valorem* duty with certainty. Instead, he threatens to impose this duty whenever the first period import price is less than the benchmark price. The discussion in chapter 4 shows that getting the Foreign exporting firm to raise its first period price is equivalent to getting the firm to certify its exports. Will this kind of threat succeed in getting the Foreign firm to undertake export certification? This is the question that chapter 4 answers.

The analysis in this chapter first establishes that the Foreign exporting firm's profit is a decreasing function of the Home *ad valorem* duty. With this result in place, chapter 4 states and solves the profit

4See Gibbons (1992, pp. 21-22) for additional details on Bertrand games.

maximization problems for the Foreign exporting and the Home firms. Analyzing the first order necessary and the second order sufficiency conditions for an optimum, chapter 4 demonstrates that there is an appropriate shift in the Foreign firm's reaction function and that this shift implies that the first period import price *goes up* from the no duty equilibrium of the Bertrand game between the two firms to the new equilibrium with the credible threat of an *ad valorem* duty by the invasive species manager in Home.

The upshot of the analysis in chapter 4 is that, if certain technical conditions involving the Foreign exporting firm's profit function and the domestic and foreign price elasticities of demand hold, then it is possible for the invasive species manager in Home to raise the first period price of the Foreign firm's export good. In turn, this is equivalent to obtaining export certification by threatening to levy a duty on the Foreign export good, if this good's first period price is less than the invasive species manager's benchmark price. Given the usefulness of an *ad valorem* duty in the strategic setting of a Bertrand game, is there a role for trade policy in general, and tariffs in particular, to serve as a deterrent against biological invasions in other market settings? This question is analyzed thoroughly in chapter 5.

2.4. Tariffs in alternate market structures

Chapter 5 makes three distinct contributions to the literature on international trade and invasive species management. First, this chapter uses a two country model to study the effectiveness of tariffs as an invasive species management tool under four different market structures. Second, the chapter studies both small and optimal tariffs. Finally, the chapter emphasizes the effects of tariffs on social welfare when social welfare depends partially on the damage from invasive species.

The two countries analyzed in the model in this chapter are called Home and Foreign. Foreign exports and Home imports a specific good. Over time, the import of this good also results in the *probabilistic* introduction of invasive species from Foreign into Home. Initially, because of scientific uncertainty, citizens and the relevant authorities in Home do not realize that these stochastically arriving invasive species cause agricultural and/or ecological damage in Home. However, over time, scientific evidence implicating the invasive species emerges and then it becomes clear to all concerned that the probabilistic introductions of these invasive species and the resulting monetary damage are *connected* to the import of the good in question from Foreign. With this realization come calls for the use of trade policy to restrict imports and thereby reduce the likelihood of the introduction of the deleterious invasive species.

The prior work of Batabyal and Beladi (2001) and Batabyal and Nijkamp (2007) is used to construct three measures of the average damage in Home from the probabilistic introduction of invasive species. Two of these three measures are monetary measures and the third is a physical measure of damage. To give the reader a flavor for the kinds of results obtained in this chapter, we now discuss the findings associated with two of the four market structures that are studied in this chapter.

The first market structure is one in which there is perfect competition in both Home and Foreign, and Home is a small country. Analysis shows that when social welfare depends on the average damage from the introduction of invasive species, small and optimal tariffs both *raise* welfare in Home, as long as the average total damage in Home is monetary. In contrast, when the average total damage is physical, a tariff is incapable of affecting social welfare in Home, and hence, in this last case, it is optimal to *not* use a tariff to regulate invasive species in Home. Even though Home is a small country, the analysis in chapter 5 shows that in two out of the three cases that are studied, tariffs have a positive impact on social welfare and hence Home ought to have an *activist* trade policy in place in these two cases.

In the last market structure studied in this chapter, there is Cournot competition between a Home firm and an exporting firm from Foreign, in the Home market. In this scenario, if a specific technical condition is satisfied, then small and optimal tariffs both generally *raise* social welfare in Home. However, in this case of Cournot competition, it is *not* inevitable that a small tariff will lead to an increase in social welfare. Specifically, if a small tariff does not raise domestic output in Home, then the effect of this small tariff on social welfare may not be positive.

When tariffs are used as delineated in this chapter, it may be possible to "kill three birds with one stone." What we mean by this expression is that the Home government may be able to obtain a terms of trade benefit, shift profits away from the Foreign exporter and toward the domestic import competing firm, and reduce the monetary damage from deleterious invasive species. This concludes our discussion of the contributions of the four chapters that focus on international trade. We now proceed to discuss the role that inspections play in efficacious invasive species management.

3. Inspections

3.1. Economic cost reduction versus bological invasion damage control

When attempting to prevent a biological invasion by inspecting the ballast water of ships, is it better for a seaport manager to have a *small* number of inspectors inspect arriving ships *less* stringently or should this manager have a *large* number of inspectors inspect the same arriving ships *more* stringently? This question is studied in a novel way in chapter 6. The focus of attention is on a stylized, publicly owned seaport, in an arbitrary country. Ships with ballast water arrive at this seaport to load cargo and to then carry this cargo to some other seaport. Occasionally, ships that come into the seaport with ballast water will first unload cargo and then load new cargo for shipment to some other seaport in the world. In either instance, the arrival of these ships coincides with the arrival of a whole host of potentially noxious biological organisms. The arrival rate of these biological organisms is proportional to the arrival rate of the ships, and hence the arrival process of the ships in the seaport represents the arrival process for the queuing theoretic inspection regimes that are analyzed in this chapter.

The manager of the seaport is interested in precluding invasions by the possibly injurious biological organisms. Therefore, arriving ships must be inspected before they can either load or unload cargo. In the first (second) model that is analyzed, the seaport has one (two) inspector(s). Hence, at any point in time, the seaport is able to simultaneously inspect either one or two ships. Ideally, the seaport manager would like the inspectors to have a zero tolerance policy. However, such a policy may be very costly to implement. Hence, chapter 6 assumes that the seaport manager is confronted with a choice between two scenarios. In the first scenario, the manager has one inspector inspect the ballast water of arriving ships not very stringently. In the second scenario, this manager has two inspectors inspect the ballast water of arriving ships more stringently.

The inspection times and the stringency of inspections are both *random* variables. Further, the inspection regime of the first scenario is *less* stringent than the inspection regime of the second scenario. Formally, the first or less stringent inspection regime is a *M/M/1* queuing model with inspection rate 2γ where γ is a positive constant and the second or more stringent inspection regime is a *M/M/2* queuing model with inspection rate γ.

Suppose that the seaport manager focuses on the "average wait of a ship in the seaport system" or *AWS* criterion. Analysis shows that *AWS* is lower when the seaport manager chooses a small number of inspectors to inspect arriving ships less stringently. In addition, if greater inspection stringency with a large number of inspectors is desired, then economic costs measured by the *AWS* criterion will be relatively high. Conversely, if the seaport manager is willing to live with less stringent inspections with a smaller number of inspectors, then *AWS* will be relatively low.

Do these findings hold when the "average wait of a ship in queue" or *AWQ* criterion is utilized by the seaport manager? New analysis shows that *AWQ* is lower when the seaport manager selects a large number of inspectors to inspect arriving ships more stringently. So, the answer to the choice of inspection regime question *does change* when the seaport manager focuses on the *AWQ* criterion. The general conclusion from chapter 6 is that there is a *tension* between economic cost reduction, on the one hand, and biological invasion damage control, on the other. If total economic cost reduction is very important, then the seaport manager will focus on the stronger *AWS* criterion and this focus means that the chosen inspection regime will be relatively less stringent. In contrast, if biological damage control is very important, then the seaport manager will select the more stringent *M/M/2* regime, and this selection is optimal only when the manager concentrates on the weaker *AWQ* criterion. Chapter 7 continues the kind of analysis conducted in chapter 6. However, the focus now is on a different managerial metric and on two different kinds of inspection policies.

3.2. Container versus time based inspection policies

Using the long run expected net cost (*LRENC*) of inspections as the apposite managerial objective, chapter 7 uses a queuing model to answer the following salient question: Given the cyclical nature of the inspection function, is the *LRENC* of inspections lower when a seaport manager follows a policy of inspecting cargo upon the arrival of a specified number of containers (the container policy) or is this *LRENC* lower when this manager inspects cargo at fixed points in time (the temporal policy)?

The chapter concentrates on a stylized, publicly owned seaport, in an arbitrary country. Upon arrival at this seaport, ships unload their containers carrying cargo and the arrival of these containers coincides with the arrival of potentially injurious biological organisms. Before this incoming cargo can be moved to interior points in the country under study, the containers must first be inspected. The purpose of this protective activity is to ensure that one or more biological invasions with potentially adverse consequences for the economy and the ecology of the pertinent country do not, in fact, take place. It is assumed that the arrival rate of the various biological organisms is proportional to the arrival rate of the containers at the inspection facility. The containers are assumed to arrive at the inspection facility in accordance with a stationary Poisson process. Clearly, the manager of the seaport is interested in precluding invasions by the possibly injurious biological organisms and, therefore, the inspection times are *random* variables. Formally, the theoretical framework of this chapter corresponds to the *M/G/1* queuing model.

The seaport manager's inspector examines the containers that have arrived at the inspection facility and he continues to do so until all the containers that are present have been inspected. When this busy period ends, the inspector leaves the inspection facility and he returns only when N new containers have arrived and are awaiting inspection. As a result of these inspections, the inspector, and ultimately the seaport manager, incurs costs and obtains benefits. Note that the inspection function is *cyclical* in nature. When containers have arrived at the facility in the seaport, the inspector is busy inspecting these containers until there are no more containers waiting to be inspected. This is the *busy* period. Then comes an *idle* period in which the inspector has no specific duties to perform. When N additional containers arrive at the facility to be inspected, a new busy period commences. This busy period is followed by an idle period, and so on and so forth. Given this state of affairs, chapter 7 uses a mathematical result known as the renewal-reward theorem—see Ross (1996, pp. 132-140) for additional details—to calculate the minimal *LRENC* when the container policy is used to conduct inspections.

Next, this chapter focuses on the temporal policy. Here, instead of selecting the number of containers optimally, the seaport manager pursues an alternate strategy. In this strategy, whenever there are no containers in the inspection facility, the inspector leaves this facility and he returns only after a fixed time period T has gone by. The task now is to choose T optimally to minimize the *LRENC* from inspection activities. Once this minimal *LRENC* has been computed, the two minimized *LRENC* functions from the use of the container and time policies are compared.

This comparison reveals that the seaport manager's *LRENC* with an optimally chosen number of containers in the inspection cycle is *lower* than his *LRENC* with an optimally chosen leave period in the inspection cycle. In other words, if the seaport manager had to choose a single control variable from a control set consisting of the number of containers and the leave period, then this manager would choose the number of containers over the leave period. Chapter 7 goes on to show that the superiority of the container policy result holds even when, instead of optimizing, the seaport manager chooses the leave period in the temporal policy on the basis of a rule of thumb. Given the demonstrated superiority of the container inspection policy, chapter 8 studies the statistical properties of what it calls a "generic" container inspection policy in some detail.

3.3. A generic container inspection policy

The environment analyzed in chapter 8 is one in which batches of containers from individual ships arrive in the seaport of a particular nation in accordance with a stationary Poisson process. The number of containers in a batch or the *batch size* is an independent random variable with a common discrete probability distribution with finite second moment. The seaport under study is large in the sense that it either actually has or can make available space to store any number of containers. This means that the arriving containers are physically stored in this seaport.

The container inspection policy studied in this chapter is generic in the sense that the inspection times of the various containers in the seaport are independent random variables with a *general* probability distribution with finite first moment. In addition, the inspection times of containers from the same batch, i.e., from the same ship, are independent of each other. The statistical properties this chapter focuses on and derives are the generating function of the batch size random variable and the limiting or stationary distribution of the number of containers present in the seaport.

The chapter points out that long delays in inspecting and clearing containers before their contents are transported to an interior point of the region in which the seaport under study is located give rise to economic costs. In addition, these economic costs are typically an increasing function of the number of containers present in the seaport. Therefore, as a practical matter, it is certainly helpful to ascertain the statistical properties of the "number of containers present" or batch size random variable. Because this number is a non-negative and integer valued random variable, it is noted that its primary statistical properties can be known by deriving the generating function associated with this number.

Chapter 8 derives this generating function and then shows that this function depends on the Poisson arrival rate of the batches of containers, the discrete probability distribution of the batch sizes, and the general probability distribution function of the inspection times of the various containers in the seaport. Following this derivation, chapter 8 determines the mean and the variance of the limiting distribution of the number of containers in the seaport under study.

The chapter points out that the mean and the variance mentioned in the previous paragraph can be used in invasive species management, in at least three ways. First, the average number of containers in the seaport can be used as an objective function by itself. When this is done, one course of action for a seaport manager might be to minimize the average number of containers by attempting to change the probability distribution function of the random inspection times and/or the value of the first moment of this distribution function. Second, the mean and the variance of the random number of containers can be part of a more general objective function that a seaport manager might want to optimize.

Finally, the mean and/or the variance can, as a constraint, be a part of an optimization problem to be solved by the seaport manager. For instance, if greater predictability in the inspection function is desired, and we are now thinking of lower variance as a sign of higher predictability, then one could formulate and solve a net social benefit from inspections maximization problem, subject to two constraints. The first constraint would tell us that the number of containers in the pertinent seaport is at most as big as an exogenously specified threshold, and the second constraint would tell us that the variance of the number of containers is also no more than a second external threshold. Clearly, the analysis conducted in chapter 8 does not exhaust the meaningful questions one can ask about inspections. Therefore, chapter 9 addresses some of these hitherto unanswered questions about inspections in the context of invasive species management.

3.4. Probabilistic features of inspection policies

Chapter 9 examines the stochastic attributes of inspections in a scenario in which there are two exporting countries and one importing country called Home. The analysis is conducted from the perspective of the manager of a representative seaport in Home. Home imports goods from two separate countries and these two exporting nations transport their goods via containers on ships to the seaport in Home. The Home seaport manager's concern is that the container cargo may contain potentially harmful invasive species Hence, this seaport manager's task is to ensure that the arriving container cargo is inspected for the presence of potentially deleterious invasive species.

Ships from the two exporting nations A and B arrive at the seaport in Home in accordance with independent and stationary Poisson processes—see Tijms (2003, pp. 1-32) for a textbook account—with rates λ_A and λ_B per time period. In addition, the time it takes to inspect the ships from A and B is exponentially distributed with means $1/\mu_A$ and $1/\mu_B$. Given this information, the chapter uses the properties of stationary Poisson processes to compute the likelihood that $n \in \mathbb{N}$ ships will need to be inspected in the next time period of interest. It turns out that the number of ships that will need to be inspected in the seaport under study in the next time period of interest is a Poisson distributed random variable.

The second task undertaken in this chapter is the determination of the probability distribution function of the amount of time it takes to inspect an *arbitrary* ship. To undertake this determination, the chapter points out that it is necessary to first determine the probability that the next ship arriving in the seaport under study is from either A or B. Because this chapter focuses on a two exporting country model, standard computations and some thought together yield a closed-form expression for the probability distribution function of the amount of time it takes to inspect an arbitrary ship. Analysis shows that this function is the weighted sum of two distinct exponential density functions where the weights are the probabilities that an arbitrary ship is either from A or from B. It is noted that this distribution function provides the Home seaport manager with a way of allocating scarce inspection resources in his seaport.

The final task undertaken in chapter 9 is to demonstrate how one might determine the probability distribution function of the *total cost* of inspections in a relevant time period. To this end, this chapter supposes that each time a ship from A (B) arrives in the Home seaport, it costs c_A (c_B) dollars to inspect this ship for the possible presence of invasive species in its container cargo. This supposition means that all ships from the same country are equally costly to inspect but ships from different countries give rise to *different* inspection costs. Analysis leads to two findings. First, it is shown that the total cost of inspections in t time units is a compound Poisson variable—see Ross (1996, pp. 82-88) or Tijms (2003, pp. 18-22) for more details—and the rate of the pertinent Poisson process is $\lambda = \lambda_A + \lambda_B$. Second, a closed-form expression for the generating function of the total cost of inspections in t time units is obtained.

Chapter 9 concludes by pointing out that the results obtained here can be used to facilitate the task of invasive species management in a variety of ways. Here are two examples. First, these results can be used to compute the expected total cost of inspections in the relevant time period of interest. When this is done, the seaport manager might select a suitable decision variable to minimize this expected total cost. Second, the likelihood that $n \in \mathbb{N}$ ships will need to be inspected in the relevant time period can play the role of a constraint in either a constrained expected total cost minimization problem or a constrained expected net social benefit maximization problem. The analyses in chapters 8 and 9 show that the *time* taken to complete inspections is a key variable in invasive species management. Therefore, chapter 10 studies container inspection policies—also see the earlier discussion about chapters 7 and 8—under alternate assumptions about the time it takes to complete inspections.

3.5. Alternate container inspection policies

Consider a stylized national point of entry, such as a seaport. Upon arrival at this seaport, ships unload their containers carrying cargo. The arrival of these containers coincides with the arrival of a whole host of potentially injurious biological organisms. Before this incoming cargo can be moved to interior points of the country under consideration, the containers must first be inspected. As in the other chapters in this third part of the book, it is assumed that the arrival rate of the various biological organisms is proportional to the arrival rate of the containers at the inspection facility. In addition, the arrival process of the containers at the inspection facility in the seaport represents the arrival process for the renewal-reward theoretic model—see Ross (1996, pp. 132-140) or Tijms (2003, pp. 39-50) for textbook treatments—that is analyzed in this chapter. The chapter assumes that the containers arrive at the inspection facility in accordance with a stationary Poisson process with time invariant rate λ and mean interarrival time 1/λ.

This chapter continues the chapter 7 inspired line of investigation and studies container inspection policies in three specific scenarios. In the first scenario the time taken to conduct inspections is negligible. In the second scenario, this inspection time is deterministic and in the third scenario this inspection time is probabilistic. Specifically, this chapter compares and contrasts the long run expected net cost or *LRENC* of inspections and the optimal container policy in each of these three scenarios.

Consider first the case of negligible inspection times. The seaport manager's inspector begins his job once N containers have arrived at the inspection facility. As a result of these inspections, the inspector (and ultimately the seaport manager) incurs costs and obtains benefits from two sources. The expected net cost per inspection cycle depends on the net costs from these two sources. The first source represents *variable* net costs and the second source denotes *fixed* net costs. Using this information, the chapter derives the *LRENC* of inspections and then minimizes this function with respect to the number of containers N. This procedure gives the optimal value of N or N_n^*. Analysis shows that as the fixed source of net cost increases, it is in the interest of the seaport manager to *raise* the optimal number of containers before commencing inspections. In contrast, if the first and variable source of net cost goes up, then the seaport manager finds it desirable to *lower* the optimal number of containers before commencing inspections.

Next, the chapter considers the case of deterministic inspection times. In this case, inspections take T time periods to complete. Once again, the chapter derives the *LRENC* from inspections and then selects the number of containers N to minimize this *LRENC* function. This gives the optimal value of N or N_d^*. It is shown that as the inspection time approaches zero, the optimal (deterministic time) container policy N_d^* converges to the optimal (negligible time) container policy N_n^*. In addition, it is also shown that, like in the case of negligible inspection times, as the fixed source of net cost increases, it is in the interest of the seaport manager to raise the optimal number of containers that must arrive before he commences inspections.

Finally, consider the case of random inspection times. The time it takes to complete inspections is now a random variable with finite first two moments denoted by $\mu_1>0$ and $\mu_2>0$ respectively. As in the previous two cases, the chapter determines the *LRENC* from inspections and then chooses the number of containers N to minimize this *LRENC* function. This yields the optimal value of N or N_p^*. Analysis leads to several specific results. Here are two examples. First, whether the

LRENC with random inspection times is larger or smaller than the *LRENC* with deterministic inspection times cannot be determined unambiguously. Second, as the first and the second moments of the random inspection time approach zero, the optimal (random time) container policy N_p^* converges to the optimal (negligible time) container policy N_n^*. Although this chapter clearly models the net costs of container inspections, it does not explicitly analyze whether there is a *tradeoff* between economic cost minimization and inspection stringency in invasive species management. This task is undertaken in chapter 11 and it is to this chapter that we now turn.

3.6. Economic cost reduction versus biological invasion damage control

Recently, DeAngelo *et al.* (2007) have studied the properties of inspections by utilizing the "average wait of a ship in the port system" or *AWS* criterion in a *M/G/1* queuing model. Their basic result is that there is *no* necessary tension between economic cost minimization and inspection stringency in invasive species management. In their paper, greater (lesser) inspection stringency reflects an enhanced (decreased) concern for the potential damage from a biological invasion. Therefore, a seaport manager who places a relatively big (small) weight on invasion damage control will, *ceteris paribus*, want to inspect ships more (less) stringently.

Chapter 11 uses an alternate cost criterion, namely, the "average wait of a ship in queue" or *AWQ* criterion to investigate the generality of the results in DeAngelo *et al.* (2007). The chapter focuses on a stylized, publicly owned seaport in some nation. Ships with ballast water and/or cargo in containers arrive at this seaport to either load or unload cargo. The arrival of these ships coincides with the arrival of a whole host of possibly damaging invasive animal and plant species. Ships arrive at the seaport in conformity with a stationary Poisson process with time independent rate $\lambda>0$. This parameter λ serves as a proxy for both the likelihood of a biological invasion and the volume of maritime trade. The analysis focuses on a *representative* dock inspector's decision problem. The inspection times are arbitrarily distributed. The seaport system under study consists of ships that are being inspected, ships that are waiting to be inspected, the representative dock inspector, and the seaport manager.

The amount of time it takes to complete inspections has a direct bearing on the stringency of inspections. To model this idea, it is assumed that there are two feasible inspection regimes in the seaport. In the first or *more* stringent inspection regime, the mean inspection time is v_M and the variance of this time is τ_M^2. In the second or *less* stringent inspection regime, the mean inspection time is v_L and the variance of this time is τ_L^2. The chapter assumes that $v_M > v_L$ and that $\tau_M^2 < \tau_L^2$.

These two inequalities *together* indicate that inspection regime M is *more* stringent than inspection regime L. Formally, the more stringent inspection regime M corresponds to a $M/G/1$ queuing model with mean inspection time v_M and variance τ_M^2 and the less stringent or L inspection regime also corresponds to a $M/G/1$ queuing model but now with mean inspection time v_L and variance τ_L^2.

Relative to the DeAngelo *et al.* (2007) paper, this chapter adopts a somewhat looser interpretation of the economic costs associated with inspections. In this interpretation, ships in *queue* are the ones that are most likely to impose economic costs on society. The corresponding cost criterion is the "average wait of a ship in queue" or AWQ criterion. The theoretical analysis in this chapter demonstrates that there may or may not be a tension between economic cost minimization and inspection stringency. Therefore, from a theoretical standpoint, even when one uses the AWQ criterion, the main finding in DeAngelo *et al.* (2007) that there is no necessary tension between economic cost minimization and inspection stringency holds.

Because of the ambiguity of the above theoretical result, chapter 11 also conducts numerical analysis and this analysis leads to two findings. First, for many values of the model parameters that describe the strictness of inspections, there *is* a tension between economic cost minimization and inspection stringency. Second, for most values of the model parameter that depicts the volume of maritime trade handled by the seaport under consideration, there is *no* tension between economic cost minimization and inspection stringency. Note that these results are different from those found in chapter 6 with the *M/M/*1 and the *M/M/*2 queuing models because of the simple reason that the queuing models employed in this chapter are *different* from the queuing models used in chapter 6.

A continuing issue with inspections is that they are generally expensive to undertake. This raises the question of alternate ways of conducting inspections when an invasive species manager must also confront the fact that he has a limited budget for inspections. One way to address this situation is to conduct inspections in a *probabilistic* manner and then levy fines on non-compliant entities. This topic is analyzed in chapter 12.

3.7. Random inspections and fines

This chapter studies a port of entry, such a seaport in an arbitrary country called Home. The analysis is conducted from the perspective of an inspection agency that has been entrusted with the task of inspecting arriving ships in this seaport for the presence of one or more invasive species. Home engages in goods trade with several nations and hence ships from these various nations arrive in the seaport in Home to unload and/or to load cargo. Now, as noted in chapter 3 and in Batabyal (2008), the risk of unintended biological invasions in Home typically varies by trading partner. Therefore, the seaport inspection agency has distinct protocols for inspecting the arriving ships from distinct nations. Put differently, ships arriving from country A are treated differently than ships arriving from country B, for any two arbitrary countries A and B.

Consider ships coming into the seaport under study from some arbitrary country, say country A. On the basis of previously collected historical data, the seaport inspection agency has determined that there is a fixed probability p that a given arriving ship from A will have one or more invasive species on it. Hence, such an arriving ship will fail to pass the agency's inspection. It is assumed that whether a particular ship from A does or does not have a problem with invasive species does not depend on the status of any other ship arriving in the Home seaport from A.

The seaport inspection agency in Home proceeds as follows. At first, it inspects every ship from A until i consecutive ships are found not to have any invasive species on them. Once this happens, the agency then inspects only one out of every r ships from A at *random* until another ship with one or more invasive species on it is discovered. When this happens, the agency reverts to one hundred percent inspections until i consecutive ships with no invasive species on them are found. The agency's inspection continues in this way. Given the above schema, the two key metrics that this chapter computes are the average fraction of all

A ships that are and are not inspected. The acronyms *AFSI* and *AFSNI* are used to refer to these two averages. The computation itself proceeds in three steps.

Analysis shows that the average fraction of arriving ships from *A* that are inspected depends on the fixed probability p that a given arriving ship will have one or more invasive species on it and on the positive integer r describing the number of ships out of which one will be inspected at random in the second stage. When either $p=1$ or $r=1$, the above described random inspection scheme is shown to be equivalent to a deterministic scheme in which all arriving ships from *A* are inspected by the agency. In addition, when $p=0$ the average fraction of ships that are inspected depends only on the positive integer r and as r increases (decreases), the average fraction of ships inspected decreases (increases). The second and final question of interest in this chapter involves the computation of the average total fines that will be collected in the long run by the inspection agency when it uses the above delineated random inspection scheme to screen arriving ships for the presence of invasive species.

Note that in addition to *A*, ships from many other countries also arrive in the seaport under study. Further, there is an inspection protocol in place for the ships from every relevant country. What happens to ships that fail the agency's inspection? In practice, agencies responsible for the management of invasive species in many countries such as Japan and the United States levy fines on non-compliant entities. Therefore, this chapter supposes that ships that fail the agency's inspection are fined and that the magnitude (dollar value) of the individual ship fines are *random* variables.

Analysis leads to a closed-form expression for the average total fines that will be collected in the long run by the inspection agency when it uses the random inspection scheme described above to screen arriving ships. This expression can be used to facilitate the task of invasive species management in two ways. First, this expression can be used to ascertain whether it is feasible to make the conduct of inspections by the agency a revenue-neutral operation. Second, the same expression can play the role of a constraint in an expected net social benefit from inspections maximization problem. Given the salience of the *probabilistic* environment in which invasive species managers operate, chapter 13—the last chapter in this third part of the book—shows how the theory of continuous-time Markov chains can be used to help us better comprehend the properties of a seaport cargo unloading and inspection protocol.

3.8. Unloading and inspection of container cargo

Chapter 13 concentrates on a seaport in a nation such as the United States that has a container terminal and a single inspector. This inspector performs two basic functions. First, he is in charge of unloading the container cargo that arrives in this seaport on ships from different parts of the world. Second, he is responsible for inspecting the container cargo he has just unloaded for the presence of one or more deleterious invasive species. The time it takes to unload the container cargo from a single ship is exponentially distributed with mean $1/\zeta_1$ where $\zeta_1>0$. The inspector then commences the task of inspecting the unloaded container cargo. This inspection task takes an exponentially distributed amount of time with mean $1/\zeta_2$ where $\zeta_2>0$. After completing the inspection task, the inspector is available to unload the container cargo from the next ship that has arrived in the seaport. The time between the arrival of successive ships in the seaport is also exponentially distributed with mean $1/\theta$ with $\theta>0$. Finally, there are a finite number of $N>0$ ships that visit the seaport under study.

The discussion in this chapter begins with two definitions. To this end, $X_1(t)$ is the number of ships that are present before the unloading inspector at any time t. Second, $X_2(t)$ is equal to one if the inspector is in the process of inspecting the unloaded container cargo from a ship at any time t and zero otherwise. Analysis shows that the stochastic process $\{X_1(t),\ X_2(t)\}$ is a continuous-time Markov chain—see Ross (1996, pp. 231-294) or Tijms (2003, pp 141-186) for textbook expositions—whose state space is given by I where $I=\{(i, j) \mid i=0,1,2,..N; j=0,1\}$.

This continuous-time Markov chain is examined in detail and this examination results in the derivation of analytic expressions for four specific inspection related metrics. The first metric is the long run fraction of time that the unloading inspector is idle. The second metric is the long run average number of ships from which cargo is unloaded per unit time. The third metric is the long run mean number of ships that are waiting to have their cargo unloaded and inspected. The fourth and final metric is the long run mean waiting time per ship in the seaport.

From a practical perspective, the first two metrics mentioned in the preceding paragraph are useful because they can be used to lower idleness in the seaport. In other words, these two metrics can be used to increase the *efficiency* of the container cargo unloading and inspection protocol that is the object of study in this chapter. The third and the fourth metrics in the previous paragraph describe alternate concepts of waiting in the seaport under study. Therefore, by minimizing one or both of these two concepts of waiting, it will, once again, be possible to positively affect protocol *performance* by lowering an aspect of the container cargo unloading and inspection protocol being studied in this chapter. This concludes our discussion of the contributions of the eight chapters that focus on inspections as a key invasive species management tool. We now proceed to discuss the contributions of the two chapters that comprise part IV of this book. These chapters focus on the nexuses between a particular kind of agro-ecosystem, namely, orchards and invasive species management.

4. Agriculture

4.1. Orchard with N trees

Chapter 14 begins by pointing out that following the work of Batabyal and Yoo (2007a, b) and others, orchards are now appropriately viewed as *jointly determined* ecological-economic systems. In addition, two points concerning orchards are worth emphasizing. First, orchards are not only jointly determined ecological-economic systems but their growth and output are also *dynamic* and *probabilistic* in nature. As such, sustainable orchard management has everything to do with the management of uncertainty over time. Second, orchards are frequently attacked by one or more invasive species such as citrus canker and the Mediterranean fruit fly. Therefore, efficient orchard management ought to take this aspect of the management problem into consideration explicitly. Unfortunately, the extant literature on orchard management has *not* satisfactorily accounted for the above two points. Given this state of affairs, chapter 14 uses renewal theory and analyzes an intertemporal and stochastic model of optimal orchard management that explicitly takes into account the damage from a successful invasion by an invasive animal or plant species.

Consider an orchard consisting of N independent and practically identical trees that is overseen by an orchard manager. As noted in the preceding paragraph, an invasive species such as citrus canker or the Mediterranean fruit fly may invade this orchard. If such an invasion occurs, then it is assumed that the trees in the orchard under study will get infected with a disease at a constant infection rate given by $\mu>0$. The orchard manager removes the infected trees from the orchard and plants new trees in their place only when the number of infected trees in this orchard reaches the critical number C, where $1 \leq C \leq N$.

The chapter supposes that this simultaneous tree removal and replanting exercise takes a negligible amount of time. The total cost of removing and replanting C trees is given by the sum of a *fixed* cost term F and a *variable* cost term αC In addition to the fixed and the variable costs, there is an additional "diseased tree" cost of β per unit time for each infected and diseased tree in the orchard before it is removed by the manager.

Next, this chapter derives a closed-form expression for the long run expected cost or *LREC* of orchard management per unit time. This *LREC* function depends on the number of trees in the orchard (N), the critical number of trees that are infected before prophylactic actions commence (C), the invasive species driven tree infection rate (μ), and the cost resulting from the presence of diseased trees in the orchard in a renewal cycle before our manager acts to remove them (β).

The orchard manager's objective is to choose the critical number of infected and diseased trees (C) that necessitate his prophylactic tree removal and replanting actions. Analysis of the pertinent minimization problem shows that the optimal critical number of trees C^* that must be infected and diseased before the manager takes his prophylactic actions ought to be chosen so that the manager equalizes the marginal impact of this choice on the weighted expected length of the relevant renewal cycle with the corresponding marginal impact on the weighted expected management cost in this same renewal cycle.

The analysis in this chapter explicitly focuses on each and every one of the N trees in the orchard. This notwithstanding, following a well established tradition in economics, it is also possible to analyze the orchard management problem by focusing the analysis on a *representative* tree in the pertinent orchard. This approach and the theory of discrete-time Markov chains are used in chapter 15, the last chapter in this fourth part of the book.

4.2. Orchard with a representative tree

This chapter begins by emphasizing the three points made in chapter 14. First, orchards are *jointly determined* ecological-economic systems. Second, sustainable orchard management has everything to do with the *management of uncertainty* over time. Third, orchards are periodically *attacked* by one or more *invasive* species. The key contention in this chapter is that even though chapter 14 and a small number of other papers have begun the process of modeling the above three points, the systematic study of the above three noteworthy points in the context of orchard management is still in its infancy.

Therefore, chapter 15 concentrates on a *representative* tree in a stylized orchard and demonstrates how the theory of discrete time Markov chains—see Ross (2003, pp. 181-268) or Tijms (2003, pp. 81-139) for textbook treatments—can be used to construct a model of this tree that is to be managed in the likely presence of one or more deleterious invasive species. In this regard, consider a fruit orchard that consists of a finite number of practically identical trees. As noted above, the focus of the orchard manager is on the representative tree in this orchard. Because all the trees in this orchard are practically identical, the full orchard management problem is essentially a scaled up version of this individual tree management problem.

At the beginning of each time period, the particular tree that the orchard manager has chosen to focus on is inspected to determine its health status. As a result of this inspection, the tree in question is deemed to be in a particular health state. The N possible health states are described with the letter i where i=1,2,…,N. In this description, health state i is better than health state i+1. Further, health state i=1 is the healthiest possible state and health state $i=N$ is the state in which the tree under study is dead, i.e., it is in the worst possible health state.

If the tree under study is in health state $i=N$, then this tree must be removed and replaced with a new tree. This *mandatory* removal and replacement activity takes two time periods. If the tree is in some intermediate health state $i<N$ then the orchard manager faces a choice between *preventively* removing and replacing this tree or letting this tree grow for the present time period. Preventive tree removal and replacement takes one time period. A new or replacement tree is uninfected and in the best possible health state.

Suppose that the pertinent tree's current health state is i. Then, as a result of possible attack by one or more invasive species, in the next time period, its health state will be j with probability q_{ij} where q_{ij}=0 for $j<i$. The orchard manager uses the following rule to determine when to preventively remove and replace the tree under consideration. This tree is removed and replaced with a new tree only when the health state of

the tree is greater than a critical value *m*, where *m* is a positive integer with the property that $1 \leq m \leq N$. The chapter next shows that the above setup can be described by a specific discrete-time Markov chain.

Once this has been done, chapter 15 computes two analytic expressions. The first expression is the long run fraction of time during which the representative tree is unhealthy. There are two implications of this long run fraction for orchard management. First, it is shown that the manager can *lower* the long run fraction of time the tree is unhealthy by "weakening" the decision rule he is using to determine when the tree under study ought to be preventively removed and replaced. Second, as *m* approaches *N*, i.e., as the manager chooses to focus *only* on mandatory tree removal and replacement, the long run fraction of time the representative tree is unhealthy is, in a probabilistic sense, as small as it can possibly be.

The second analytic expression is the long run fraction of tree replacements that occur when the representative tree under consideration is dead. Now, *ceteris paribus*, the orchard manager will want to *minimize* this second analytic expression. This minimization can be undertaken by "strengthening" the decision rule used by the manager to determine when he should preventively remove and replace the tree under study. From a practical perspective, this minimization will involve *lowering* the value of *m* so that this critical health state is now closer to the best health state $i=1$. Further, as m approaches 1, i.e., as the manager chooses to focus on preventive tree removal and replacement, the long run fraction of tree replacements that are mandatory will, in a probabilistic sense, be as small as it can possibly be. This completes our discussion of the two chapters comprising part IV of this book.

5. Conclusions

The work of Pimentel *et al.* (2000) and Pimentel *et al.* (2005) clearly indicates that the unintentional introductions of invasive species have frequently resulted in significant ecological and economic damages. For the United States alone, these researchers have estimated the total monetary damage from invasive species (in 2003 US dollars) in 2005 to be approximately $128 billion. Similar monetary damage measures for many other countries can be found in Lovell *et al.* (2006) and in Olson (2006). This state of affairs tells us that in this era of globalization in which the transportation of humans, goods, and services is occurring at unprecedented rates, the management of invasive species is a task of considerable salience.

This task is made difficult by the fact that an invasive species manager has to account for several sources of uncertainty in determining his or her optimal course of action in a given situation. The 14 chapters of this book that follow this introductory chapter use modern methods from economics and operations research to model and study effective decision making under uncertainty. In this way, these chapters collectively demonstrate how different facets of this uncertainty can be understood and acted upon. The rigorous, theoretical study of invasive species management is a relatively recent endeavor. Therefore, in the coming years, one may look forward to many thought provoking developments in theoretical research in this burgeoning new field of inquiry.

References

Acquaye, A.K.A., Alston, J.A., Lee, H., and Sumner, D.A. 2005. Economic consequences of invasive species policies in the presence of commodity programs: Theory and application to citrus canker, *Review of Agricultural Economics*, 27, 498-504.

Batabyal, A.A. 2004. A research agenda for the study of invasive species introduced unintentionally via maritime trade, *Journal of Economic Research*, 9, 191-216.

Batabyal, A.A. 2008. Two theoretical issues concerning the differential treatment of trading partners in alien species management, *Ecological Economics*, 64, 679-682.

Batabyal, A.A., and Beladi, H. 2001. Aspects of the theory of financial risk management for natural disasters, *Applied Mathematics Letters*, 14, 875-880.

Batabyal, A.A., and Nijkamp, P. 2007. The stochastic arrival of alien species and the number of and the damage from biological invasions, *Ecological Economics*, 62, 277-280.

Batabyal, A.A., and Yoo, S.J. 2007a. A probabilistic approach to optimal orchard management, *Ecological Economics*, 60, 483-486.

Batabyal, A.A., and Yoo, S.J. 2007b. A stochastic analysis of the Holling resilience of an orchard, *Ecological Economics*, 64, 1-4.

Brown, R. 2006. Exotic pets invade United States ecosystems: Legislative failure and a proposed solution, *Indiana Law Journal*, 81, 713-731.

DeAngelo, G., Batabyal, A.A., and Kumar, S. 2007. On economic cost minimization versus biological invasion damage control, in A. Oude Lansink, (Ed.), *New Approaches to the Economics of Plant Health*. Springer, Heidelberg, Germany.

Gibbons, R. 1992. *A Primer in Game Theory*. Harvester Wheatsheaf, Hempstead, UK.

Horan, R.D., Perrings, C., Lupi, F., and Bulte, E.H. 2002. Biological pollution prevention strategies under ignorance: The case of invasive species, *American Journal of Agricultural Economics*, 84, 1303-1310.

Lovell, S.J., Stone, S.F., and Fernandez, L. 2006. The economic impacts of aquatic invasive species: A review of the literature, *Agricultural and Resource Economics Review*, 35, 195-208.

McAusland, C., and Costello, C. 2004. Avoiding invasives: Trade-related policies for controlling unintentional exotic species introductions, *Journal of Environmental Economics and Management*, 48, 954-977.

McConnachie, A.J., de Wit, M.P., Hill, M.P., and Byrne, M.J. 2003. Economic evaluation of the successful biological control of *Azolla filiculoides* in South Africa, *Biological Control*, 28, 25-32.

Mooney, H.A. 1999. Species without frontiers, *Nature*, 397, February 25, 665-666.

Nogales, M., Martin, A., Tershy, B.R., Donlan, C.J., Veitch, D., Puerta, N., Wood, B., and Alonso, J. 2004. A review of feral cat eradication on islands, *Conservation Biology*, 18, 310-319.

Olson, L.J. 2006. The economics of terrestrial invasive species: A review of the literature, *Agricultural and Resource Economics Review*, 35, 178-194.

Perrings, C., Williamson, M., and Dalmazzone, S. Eds. 2000. *The Economics of Biological Invasions*. Edward Elgar, Cheltenham, UK.

Pimentel, D., Lach, L., Zuniga, R., and Morrison, D. 2000. Environmental and economic costs of nonindigenous species in the United States, *BioScience*, 50, 53-65.

Pimentel, D., Zuniga, R., and Morrison, D. 2005. Update on the environmental and economic costs associated with alien-invasive species in the United States, *Ecological Economics*, 52, 273-288.

Roberts, D. 1999. Analyzing technical trade barriers to agricultural markets: Challenges and priorities, *Agribusiness*, 15, 335-354.

Ross, S.M. 1996. *Stochastic Processes*, 2nd edition. Wiley, New York, NY.

Ross, S.M. 2003. *Introduction to Probability Models*, 8th edition. Academic Press, San Diego, CA.

Tijms, H.C. 2003. *A First Course in Stochastic Models*. Wiley, Chichester, UK.

Vitousek, P.M., D'Antonio, C.M., Loope, L.L., Rejmanek, M., and Westbrooks, R. 1997. Introduced species: A significant component of human-caused global change, *New Zealand Journal of Ecology*, 21, 1-16

CHAPTER 2

INTERNATIONAL TRADE AND BIOLOGICAL INVASIONS: A QUEUING THEORETIC ANALYSIS OF THE PREVENTION PROBLEM - WITH H. BELADI

Abstract. We propose and develop a new framework for studying the problem of preventing biological invasions caused by ships transporting internationally traded goods between countries and continents. Specifically, we apply the methods of queuing theory to analyze the problem of preventing a biological invasion from a long run perspective. First, we characterize two simple regulatory regimes as two different kinds of queues. Second, we show how to pose a publicly owned port manager's decision problem as an optimization problem, using queuing theoretic techniques. Third, we compare and contrast the optimality conditions emanating from our analysis of the *M/M/I/U* and the *M/M/I/I* inspection regimes. We conclude by discussing possible extensions to our basic models.

1. Introduction

In this age of globalization, there is increasing mobility of both humans and goods between countries and continents. Ships are routinely used to transport a variety of internationally traded goods between different countries. There is no gainsaying the fact that this international trade in *goods* is generally beneficial to the countries involved. Indeed, there are several results in modern trade theory which show that voluntary goods trade between nations is welfare improving for all the nations involved.

This notwithstanding, as Heywood (1995), Parker *et al.* (1999), and others have pointed out, in addition to transporting goods between countries, by means of their ballast water, ships have also unwittingly transported all kinds of non-native—also referred to as exotic or invasive—plant and animal species from one geographical region to another.[5] These non-native or alien species have often been very successful in invading their new habitats and the resulting biological invasions have proved to be very costly to the countries in which these new habitats are located. For the United States alone, the magnitude of these costs is astounding. For instance, according to the Office of Technology Assessment (OTA (1993)), the Russian wheat aphid caused an estimated $600 million worth of crop damage between 1987 and

5The principal method of marine non-native species introduction is by means of the dumping of ballast water. Cargo ships typically carry ballast water in order to enhance maneuverability and stability when they are not carrying full loads. When these ships come into port, this ballast water must be discharged before cargo can be loaded. It is estimated that over 4000 species of invertebrates, algae, and fish are being moved around the world in ship ballast tanks every day. Focusing on just one country, it has been estimated that as much as 13 billion gallons or 50 million metric tonnes of overseas ballast water enters Canadian coastal ports every year. A recent study by the Smithsonian Environmental Research Center (SERC) in Edgewater, Maryland calculated that a liter of ballast water typically contains several billion organisms similar to viruses and up to 800 million bacteria. For more details on these issues, go to http://www.fundyforum.com/profile_archives and to the SERC web site www.serc.si.edu

1989. More generally, Pimentel *et al.* (2000) have estimated the total costs of all non-native species to be around $137 billion per year.

It is important to understand that, in addition to economic costs, invasive species also cause significant ecological damage. As Vitousek *et al.* (1996) and de Wit *et al.* (2001) have noted, non-native species can alter ecosystem processes, act as vectors of diseases, and diminish biological diversity. In this regard, the work of Cox (1993) tells us that out of 256 vertebrate extinctions with a known cause, 109 are the result of biological invasions. Even a single invasive species can cause tremendous damage. Savidge (1987) tells us that following an invasion of Guam by the brown tree snake, all twelve of this island's bird species became extinct.

The point of this discussion is clear. Biological invasions can be and frequently have been a huge menace to society. Given this state of affairs, one can ask what economists have contributed to increasing our understanding of the regulation of biological invasions. Unfortunately, the answer is not much. Although, very recently, economists have begun to address this question, it is still the case that "the economics of the problem has...attracted little attention" (Perrings *et al.* (2000, p. 11)).

From a regulatory perspective, there are a number of actions that one can take to deal with the problem of biological invasions. It is helpful to separate these actions into pre-invasion and post-invasion actions. Pre-invasion actions relate to the so called prevention problem. The idea here is to take actions that will effectively *prevent* a potentially damaging non-native species from invading a new habitat. In contrast, post-invasion actions involve the optimal control of one or more non-native species, given that the species has already invaded a new habitat.

Most economic analyses of the regulation of biological invasions have focused on the desirability of alternate actions in the *post-invasion* scenario. We now briefly discuss four representative studies. Barbier (2001) shows that the economic impact of a biological invasion can be determined by studying the nature of the interaction between the non-native and the native species. He notes that the economic impact depends on whether this interaction involves interspecific competition or dispersion. Eiswerth and Johnson (2002) analyze an optimal control model of the management of a non-native species stock. They show that given presently available scientific information, the optimal level of management effort is sensitive to ecological factors that are species- and site-specific and stochastic. Olson and Roy (2002) have used a model of a stochastic biological invasion to examine conditions under which it is optimal to eradicate the non-native species and conditions under which it is not optimal to do so. Finally, Eiswerth and van Kooten (2002) have shown that, even when hard data about the spread of an invasive species are unavailable, it is possible to use information provided by experts to formulate a model in which it is optimal to not eradicate but instead control the spread of an invasive species.

The above studies have certainly increased our understanding of regulatory issues in the *post-invasion* scenario. This notwithstanding, to the best of our knowledge, the only paper that has formally analyzed the prevention problem, i.e., the regulation of a potentially damaging non-native species before invasion is Horan *et al.* (2002). These researchers model non-native invasive species as a form of "biological pollution." They then compare the properties of preventive management strategies under full information and under uncertainty. This chapter is different from the above paper in three important ways. First, we are not interested in comparing the properties of management strategies under full information and under uncertainty. In this regard, we suppose from the beginning that uncertainty is an integral component of the prevention problem confronting a regulator. Second, we use queuing theory—to the best of our knowledge for the first time—to provide a *long run* perspective on the stochastic setting in which our regulator operates. Finally, we use aspects of this stochastic setting to set up objective functions that our regulator optimizes.

The rest of this chapter is organized as follows: Section 2 first provides a brief primer on queuing theory and then it focuses on the two queuing models that we use to study the prevention problem confronting our regulator. Section 3 uses the first of these two queuing models to provide a detailed analysis of the regulator's prevention problem. Section 4 does the same using the second of our two queuing models. Section 5 compares and contrasts the optimality conditions emanating from our analysis of the two specific queuing inspection regimes. Section 6 concludes and offers suggestions for future research.

2. Queuing Theory and the Prevention Problem

2.1. A primer on queuing theory

Queuing theory is concerned with the mathematical analysis of waiting lines or queues.[6] At a very elementary level, all queuing models have three characteristics. In particular, they can be described by (i) a stochastic arrival process, (ii) a random service time (times) distribution function (functions), and (iii) the deterministic number of available servers. The arrival process is often but not always described by the Poisson process. When this is the case, the times between successive arrivals are exponentially distributed and, as is well known, the exponential distribution is memoryless or Markovian in nature. Consequently, the Poisson arrival process is commonly described by the letter *M*.

6Excellent textbook accounts of queuing theory can be found in Gross and Harris (1974), Wolff (1989), and Ross (2003).

The service times are clearly stochastic and hence these times can, in principle, be arbitrarily distributed. However, these service times are frequently modeled with the exponential distribution function which is memoryless or Markovian in nature. Hence, in this case, the letter M is also used to represent the service time distribution function. Finally, the deterministic number of servers is typically denoted by some positive integer.

So, for instance, the notation $M/M/1$ refers to a queuing model in which the arrival process is Poisson, the service time is exponentially distributed, and there is a single server. Similarly, the notation $M/G/5$ refers to a queuing model in which the arrival process is Poisson, the service times are generally distributed, and there are five servers. It is possible to complicate this basic three-part construct in several ways, and in this chapter we shall do so by adapting this basic three-part construct to our biological invasion prevention problem.

2.2. The biological invasion prevention problem

Consider a stylized, publicly owned port in a specific coastal region of some country. Ships with ballast water arrive at this port typically to load cargo and to then transport this cargo to some other port. Occasionally, it may also be the case that some ships that come into our port with ballast water will first unload cargo and then load new cargo for shipment to some other port in the world. In either case, the arrival of these ships coincides with the arrival of a whole host of (potentially deleterious) biological organisms. It is reasonable to suppose that the arrival rate of these biological organisms is proportional to the arrival rate of the aforementioned ships. Therefore, we shall not model these biological organisms directly. Instead we shall focus on the ships that bring—by means of their ballast water—these organisms to our port. Given this interpretation, the arrival process of the ships in our port constitutes the arrival process for the queuing models that we employ in this chapter. Now, consistent with a wide variety of queuing models, we suppose that the ships in question arrive at our port in accordance with a Poisson process with rate α.

Because we are interested in preventing invasions by the potentially deleterious biological organisms, arriving ships must be inspected, before they can either load or unload cargo. We assume that our port has I inspectors, where I is some positive integer. Put differently, at any point in time, our port will be able to simultaneously inspect a maximum of I arriving ships. Further, ships are inspected on a first come, first served basis. If more than I ships arrive at our port during a particular time interval, then the ships that are not already being inspected must wait in queue. An alternate interpretation of this state of affairs is that our port has I docks and that one inspector is assigned to each of these I docks. Therefore, at any specific moment in time, a maximum of I ships can be docked and inspected. Finally, since no port is physically able to accommodate an arbitrarily large number of ships, we suppose that there is an upper limit U on the maximum number of ships that can be allowed to queue in our port. The port system consists of ships that are being inspected, ships that are waiting in queue, the I inspectors, and the port manager.

Note that, because we are studying the prevention of biological invasions in this chapter, the inspectors in our port will typically want to make the likelihood of a biological invasion as small as possible. However, despite the best attempts of our inspectors, it is possible that there will be some slippage in the combined inspection activities in our port system. Put differently, even though our inspectors take concrete steps to reduce the chance of a biological invasion, they typically do not work with a zero tolerance policy. Why not? This is because a zero tolerance policy is generally extremely costly to implement. From this discussion, it follows that inspection is necessarily a laborious and time consuming activity.

Before a ship can be cleared for loading or unloading cargo, an inspector must have carefully examined this ship's ballast water for potentially harmful organisms. As a part of this careful examination, a specific inspector can take a number of actions. These include (i) the shipboard filtration of ballast water, (ii) the treatment of ballast water with heat, chemicals, and ultraviolet radiation, and (iii) the shore-based treatment of ballast water. In this connection, the reader should note two points. First, each of the above mentioned inspection activities can be conducted at varying levels of effectiveness. For instance, ballast tanks can be subjected to heat treatment of differing levels of intensity. Second, inspectors will not always inspect for actual biological organisms—which may be too small to detect in large quantities of ballast water. On some occasions, it may make more sense to look at what preventive technologies ships have used in transit to destroy potentially injurious biological organisms.

Given the discussion in the previous two paragraphs, it is clear that inspections will typically require varying amounts of time. For instance, if an inspector knows that a particular ship has taken on ballast water in an area where there are no known biological invaders, then (s)he may be able to clear a ship relatively quickly. In contrast, if it is the case that a particular ship has taken on ballast water during a phytoplankton bloom, then the chance of this ship's ballast water containing potentially detrimental organisms is much higher, and hence a lot more time will be required to clear this ship. This discussion should convince the reader that the time taken to complete an inspection is necessarily a *random* variable. Therefore, we suppose that this random variable is exponentially distributed with mean $1/\beta$.

We now have all the essential components of our queuing models. We shall analyze two specific inspection regimes. In both regimes, the arrival process of ships is Poisson with rate α and the inspection times are exponentially distributed with mean $1/\beta$. In the first inspection regime, there are I inspectors and the upper limit on the maximum number of ships in our port is U. In the second inspection regime, the number of inspectors and the upper limit on the maximum number of ships in our port coincide and they are both denoted by the positive integer I. Using the language of queuing theory, our first inspection regime is a $M/M/I/U$ model and our second inspection regime is a $M/M/I/I$ model. In this notation, the meaning of the first two $M's$ has already been explained in the last paragraph of section 2.1. In addition, from the above explanation, it should be clear that the I refers to the number of inspectors and the U refers to the finite capacity of our port. We now proceed to a formal discussion of our queuing theoretic approach to the biological invasion prevention problem.

3. The *M/M/I/U* Inspection Regime

3.1. The probabilistic essentials

Recall that our analysis of the prevention problem is being conducted from a long run perspective. As such, our first task is to determine the long run or stationary probabilities for our $M/M/I/U$

inspection regime. To this end, let $X(t)$ denote the number of ships in our port at an arbitrary time t. Further, let

$$P_k = \lim_{t\to\infty} Prob\{X(t)=k\} \tag{1}$$

denote the steady state or stationary probability that there are exactly k ships in our port. We are interested in determining the $\{P_k\}$. However, before we do this, it is important to note two things. First, in the queuing models of this chapter, P_k can also be interpreted as the proportion of *time* that the port system contains exactly k ships. Second, because the finite capacity of our port is U, the state space of this inspection regime can be indexed by k where k runs from 0 to U. In words, this means that when there are U ships in the port, no additional ships will be permitted to enter this port.

To compute the $\{P_k\}$, note that because of the upper limit U on the maximum number of ships that may enter our port, the relevant arrival rate of ships is not α but

$$\alpha_k = \begin{cases} \alpha & if\ 0 \le k < U \\ 0 & if\ k \ge U. \end{cases} \tag{2}$$

Similarly, because of the presence of this finite capacity, the pertinent inspection rate is also not β but

$$\beta_k = \begin{cases} k\beta \text{ } if \, 0 \leq k < I \\ I\beta \text{ } if \, I \leq k \leq U. \end{cases} \tag{3}$$

Now, the correct expression for P_k will depend on whether the actual number of ships in our port (which are indexed by k) satisfies the condition ($0 \leq k < I$) or the condition ($I \leq k \leq U$) The first condition says that the actual number of ships is less than the total number of inspectors I and, hence, at any given point in time, some inspectors are idle. The second condition says that the actual number of ships lies somewhere in between the total number of inspectors and the finite capacity of the port U. Now, the threat of a biological invasion is greatest in ports where there is a lot of ballast water that needs to be inspected. In turn, there will be a lot of ballast water when a number of ships with ballast water arrive at our port. In this case, the actual number of ships will, most likely, exceed the total number of inspectors and, hence, inspectors are unlikely to be idle. The upshot of this discussion is that, given the subject matter of this chapter, the more interesting and the more realistic of the two conditions is the condition ($I \leq k \leq U$). Therefore, in the rest of this section, we suppose that the condition ($I \leq k \leq U$) holds. Using this condition, equations (2)-(3), and equations 3.24 and 3.25 in Gross and Harris (1974, p. 105), we can tell that the required stationary probability P_k satisfies

$$P_k = \left(\frac{1}{I^{k-I} I!}\right)\left(\frac{\alpha}{\beta}\right)^k \left[\sum_{k-0}^{k=I-1} \frac{1}{k!}\left(\frac{\alpha}{\beta}\right)^k + \frac{(\alpha/\beta)^I}{I!}\left(\frac{1-\rho^{U-I+1}}{1-\rho}\right)\right]^{-1}, k=I,\ldots,U, \quad (4)$$

where $\rho=\alpha/I\beta$.[7] This completes our first main task.

The reader should note that the arrival of ships into our port does not result only in the arrival of potentially damaging biological invasions. Specifically, the loading and the unloading of cargo in our port constitutes economic activity driven by international trade between our port and ports in other nations. This trade driven economic activity clearly results in benefits to society and hence any reasonable analysis of the biological invasion prevention problem must account for this positive impact of economic activity on society. If we suppose that the volume of trade driven economic activity is proportional to the number of ships S in our port, then the *expected* number of ships $E[S]$ can serve as a useful proxy for the magnitude of this trade driven economic activity. Consequently, our next task is to determine $E[S]$ for our $M/M/I/U$ inspection regime.

7We are assuming here that $\rho \neq 1$. If this condition does not hold then the expression for P_k in equation (4) will be a little different. For additional details on this point, see Gross and Harris (1974, p. 105).

The reader should note that the expectation $E[S]$ is actually the sum of two parts. The first part is the expected number of ships that are in queue, waiting to be inspected, and the second part is the expected inspection time. Using this fact and equations 3.26 and 3.27 in Gross and Harris (1974, pp. 106-107), we reason that, in our model, $E[S]$ is given by

$$E[S]=\frac{P_0(I\rho)^I\rho}{I!(1-\rho)^2}[1-\rho^{U-I+1}-(1-\rho)(U-I+1)\rho^{U-I}]$$

$$+I-P_0\sum_{k=0}^{k=I-1}\frac{(I-k)(\rho I)^k}{k!}, \tag{5}$$

where $P_0=[\sum_{k=0}^{k=I-1}(1/k!)(\alpha/\beta)^k+\{(\alpha/\beta)^I/I!\}\{(1-\rho^{U-I+1})/(1-\rho)\}]^{-1}$.

This completes our discourse on the probabilistic essentials. We now follow Batabyal (1996) and first formulate and then discuss an optimization problem that describes the prevention problem confronting the manager of our publicly owned port.

3.2. The optimization problem

Our port manager understands that the international trade driven economic activity in this port coupled with the need for inspections to keep out potentially deleterious biological organisms generates benefits and costs to society. Therefore, our port manager is interested in optimizing the net benefit to society and this net benefit is given by the gross benefit resulting from international trade driven economic activity less the cost of preventing biological invasions.

Let us consider the benefits from economic activity, first. Ships arrive in our port at the rate α. However, a certain proportion of these ships, i.e., those that arrive when there are U ships already in the port, do not enter this port. Now, P_U is the proportion of time that our port is full. From this it follows that in a particular time period, say a month, entering ships in effect arrive at our port at the rate of $\alpha(1-P_U)$.[8] Let the per month benefit to society from the trade driven economic activity resulting from the arrival of the *kth* ship be $B_k = B_k(E[S], 1/\beta, t_k)$ where $E[S]$ is the expected number of ships in the port, $1/\beta$ is the average time taken by an inspector to inspect a ship, and t_k is the total tonnage of the goods being loaded and/or unloaded from the *kth* ship. Given this specification of the individual ship benefit function, we can see that the aggregate benefit facing our port manager per month equals $\alpha(1-P_U)$

8From an intuitive standpoint, the reader may find it helpful to think of the arrival rate of ships in terms of a positive integer. Note, however, that if necessary, this feature can easily be accounted for in our modeling framework by positing that ships arrive at the rate of the integer part of $\alpha(1-P_U)$. We suppose the reader understands this. Therefore, in the rest of this chapter, we shall not dwell on this detail.

$\sum_{k=1}^{k=a(1-P_U)} B_k(E[S], 1/\beta, t_k)$. Moving to the costs, we suppose that the per month cost of preventing biological invasions depends on the expected number of ships in our port system $E[S]$, on the mean inspection time $1/\beta$ and on the number of inspectors I who are working in this port. As such, this per month total cost can be expressed as $C(E[S], 1/\beta, I)$.

The reader should note three features of our benefit and cost modeling, thus far. First, both the total benefit and cost functions depend on a system aggregate, i.e., on $E[S]$, and on the mean inspection time $1/\beta$. Second, the only difference between the aggregate benefit and cost functions lies in the third argument in each of these two functions. In particular, this third argument is t_k (the total tonnage being loaded/unloaded from the *kth* ship) in the aggregate benefit function and it is I (the total number of inspectors) in the aggregate cost function. Finally, we have assumed that the aggregate benefit and cost functions depend on the expected number of ships $E[S]$ and *not* on the actual number of ships S. Recall that S is a random variable and, hence, if we were to make the aggregate benefit and cost functions depend on S directly, then these aggregate benefits and costs would themselves be *random*. However, this randomness in the aggregate benefits and costs at all points in time, and hence in the steady state as well, does not square very well with one of our stated objectives, i.e., to provide a long run or steady state analysis of the biological invasion prevention problem.

Therefore, we have "smoothed over" the above mentioned randomness in the aggregate benefits and costs by supposing that the steady state number of ships can be proxied well by the expected number of ships.[9]

Now, using the delineation of benefits and costs from the previous two paragraphs, we can state our publicly (and not privately) owned port manager's optimization problem. This manager chooses the number of inspectors I and the inspection rate β to solve[10]

$$\max_{\{I,\beta\}} \alpha(1-P_U) \sum_{k=1}^{k=a(1-P_U)} B_k(E[S], 1/\beta, t_k) - C(E[S], 1/\beta, I). \qquad (6)$$

Now, suppose that the solution to the problem in (6) yields an interior maximum. Then, omitting the complementary slackness conditions, the Kuhn-Tucker conditions for a maximum are

---

9As an alternative, one could conceivably work with stochastic processes of aggregate benefits and costs that converge to some long run or steady state function. We leave this alternate approach for subsequent research.

10The reader will note that we are implicitly treating I as a continuous choice variable. Put differently, we are assuming that the optimal integer I can be approximated well by the optimal continuous I. If this is not the case, then integer programming techniques will have to be used to determine the optimal number of inspectors. For more on integer programming, see Wolsey (1998).

$$\alpha(1-P_U)\sum_{\forall k}\{\frac{\partial B_k(\bullet)}{\partial E[S]}\frac{\partial E[S]}{\partial I}\}-[\alpha^2(1-P_U)\,B_{\alpha(1-P_U)}(\bullet)+\sum_{\forall k}B_k(\bullet)]\frac{\partial P_U}{\partial I}$$

$$=\frac{\partial C(\bullet)\partial E[S]}{\partial E[S]\;\partial I}+\frac{\partial C(\bullet)}{\partial I} \tag{7}$$

and

$$\alpha(1-P_U)\sum_{\forall k}\{\frac{\partial B_k(\bullet)}{\partial E[S]}\frac{\partial E[S]}{\partial \beta}-\frac{\partial B_k(\bullet)}{\partial(1/\beta)}\frac{1}{\beta^2}\}-[\alpha^2(1-P_U)\,B_{\alpha(1-P_U)}(\bullet)+\alpha$$

$$\sum_{\forall k}B_k(\bullet)]\frac{\partial P_U}{\partial \beta}=\frac{\partial C(\bullet)\partial E[S]}{\partial E[S]\,\partial \beta}-\frac{\partial C(\bullet)}{\partial(1/\beta)}\frac{1}{\beta^2}. \tag{8}$$

The optimal number of inspectors I^* and the optimal inspection rate β^* solve equations (7) and (8) respectively. Together, these two equations reveal the basic tradeoff confronting our port manager. Let us examine each of these two first order necessary conditions in greater detail. Equation (7) tells us that in selecting the number of inspectors optimally, the port manager will equate the marginal benefit from economic activities (the LHS) with the marginal cost of preventing biological invasions (the RHS).

Examining the LHS of equation (7) in greater detail, we see that the marginal benefit from economic activities is actually the weighted difference of two terms. The weight on the first term $\alpha(1-P_U)\sum_{\forall k}\{\partial B_k(\bullet)/\partial E[S]\}\{\partial E[S]/\partial I\}$ is $\alpha(1-P_U)$, the *effective* arrival rate of ships in our port. Further, this first term captures the *indirect* impact that the optimal number of inspectors has on the marginal benefit through the $E[S]$ variable, i.e., the expected number of ships in our port. The weight on the second term $[\alpha^2(1-P_U)B_{\alpha(1-P_U)}(\bullet)+\sum_{\forall k}B_k(\bullet)]\{\partial P_U/\partial I\}$ is $[\alpha^2(1-P_U)B_{\alpha(1-P_U)}(\bullet)+\sum_{\forall k}B_k(\bullet)]$, a complicated function of the individual ship benefit functions. This second term captures the *direct* effect that the optimal number of inspectors has on the marginal benefit through the stationary probability P_U that there are a maximum of U ships with ballast water in the port under study.

Moving to the RHS of equation (7), we see that the marginal cost of preventing biological invasions is the *sum* of two terms. The first term $\{\partial C(\bullet)/\partial E[S]\}\{\partial E[S]/\partial I\}$ captures the *indirect* effect that the optimal number of inspectors has on the marginal cost through the $E[S]$ variable. The second term $\{\partial C(\bullet)/\partial I\}$ accounts for the *direct* impact that the optimal number of inspectors has on the marginal cost of biological invasions. Since equation (7) cannot, in general, be solved analytically, one will have to resort to numerical methods to ascertain the optimal number of inspectors I^*. One useful way to think of the maximization problem in (6) and the corresponding first order necessary condition (7) is as follows. The port manager is solving a long run net benefit maximization problem. In the present section's version of this problem, the long run port capacity is fixed at U and our manager's task is to endogenously select the optimal number of inspectors. If the optimal number of inspectors I^* exceeds the fixed port capacity U, then to avoid the problem of idle inspectors, our port manager will presumably want to set $I^*=U$.

Equation (8) gives us the optimal inspection rate β^*. The interpretation of this equation is similar to that of equation (7) and hence we shall be relatively brief. As before, our port manager chooses the inspection rate so as to equate the marginal benefit from inspection at the optimal rate with the corresponding marginal cost. Specifically, the marginal benefit (the LHS of equation (8)) is the weighted difference of two terms. The first term captures the *indirect* impact that the optimal inspection rate β^* has on the marginal benefit through the $E[S]$ variable and through the mean inspection time $1/\beta$. The second term on the LHS of equation (8) captures the *direct* effect that the optimal inspection rate β^* has on the marginal benefit through the steady state probability P_U that there are a maximum of U ships with ballast water in the port under study.

Proceeding to the RHS of equation (8), we see that the marginal cost of preventing biological invasions is the *difference* of two terms. The first term $(\partial C(\bullet)/\partial E[S]\}\{\partial E[S]/\partial \beta\}$ captures the impact that the optimal inspection rate β^* has on the marginal cost through the $E[S]$ variable. The second term $(\partial C(\bullet)/\partial(1/\beta)\}\{1/\beta^2\}$ accounts for the effect that the optimal inspection rate β^* has on the marginal cost of preventing biological invasions through the mean inspection time $1/\beta$. As was the case with equation (7), equation (8) too cannot, in general, be solved analytically. Hence, it will once again be necessary to use numerical techniques to determine the optimal inspection rate β^*. We now proceed to discuss the second of our two inspection regimes.

4. The *M/M/I/I* Inspection Regime

4.1. The probabilistic essentials

Unlike the *M/M/I/U* inspection regime, in the regulatory regime of this section, the number of inspectors in the port is equal to the maximum number of ships that this port can accommodate. In symbols, *I*=*U*. Now, given that our analysis is being conducted from a long run perspective, our immediate task is to compute the stationary probabilities $\{P_k\}$—given by equation (1)—for this *M/M/I/I* inspection regime. Before we calculate the $\{P_k\}$, observe that in this section the condition

($0 \le k \le I$) applies. This condition tells us that the actual number of ships in our port is less than or equal to the total number of inspectors I and, hence, at any given point in time it is possible that some inspectors are idle. The reader will note that because $I=U$, it now makes sense to work with this condition ($0 \le k \le I$) and not the condition ($I \le k \le U$) of the previous section. Now, using equation 3.31 in Gross and Harris (1974, p. 109), we infer that the long run probability P_k we seek is given by

$$P_k = \frac{(\alpha/\beta)^k / k!}{\sum_{j=0}^{j=I} (\alpha/\beta)^j / j!}, \; k=0,\ldots,I. \tag{9}$$

This completes our first primary task.

As in section 3, the loading and the unloading of cargo in our port represents economic activity driven by international trade between our port and ports in other countries. This trade driven economic activity obviously results in benefits to society and hence we shall account for these benefits in our analysis. To this end, we assume that the *expected* number of ships $E[S]$ is a useful proxy for the magnitude of this trade driven economic activity. As such, our next task is to ascertain $E[S]$ for the *M/M/I/I* inspection regime. To obtain the relevant $E[S]$, let us substitute $I=U$ in equation (5) and then simplify the resulting expression. This gives us

$$E[S] = I - P_0 \sum_{k=0}^{k=I-1} \frac{(I-k)(\rho I)^k}{k!}, \tag{10}$$

where $P_0 = [\sum_{k=0}^{k=I-1} (1/k!)(\alpha/\beta)^k + \{(\alpha/\beta)^I / I!\}]^{-1}$.

This completes our discussion of the probabilistic essentials. We now formulate and then discuss an optimization problem that characterizes the prevention problem facing the manager of our publicly (and not privately) owned port.

4.2. The optimization problem

Our port manager is aware of the fact that the international trade driven economic activity in this port, combined with the need for inspections to keep out possibly injurious biological organisms, results in benefits and costs to society. Therefore, this port manager is interested in maximizing the net benefit to society and this net benefit is given by the gross benefit resulting from international trade driven economic activity less the cost of preventing biological invasions.

Let us consider the benefits from economic activity, first. Following the discussion in section 3.2, the gross benefit facing our port manager per month is $\alpha(1-P_I) \sum_{k=1}^{k=\alpha(1-P_I)} B_k(E[S], 1/\beta, t_k)$, where the arguments of this benefit function are as in section 3.2. The reader will note that this benefit function is different from the benefit function of the *M/M/I/U* inspection regime of section 3 in three ways. First, the effective arrival rate is $\alpha(1-P_I)$ and not $\alpha(1-P_U)$. Second, the upper limit of the summation now is $\alpha(1-P_I)$ and not $\alpha(1-P_U)$ Finally, as equations (5) and (10) tell us, the expression for the $E[S]$ argument in the above benefit function is not the same as the corresponding expression for the section 3.2 benefit function.

As far as the costs are concerned, we follow the logic of section 3.2 and suppose that the per month cost of preventing biological invasions is a function of the expected number of ships in our port $E[S]$ the mean inspection time 1β, and the number of inspectors I who are working in this port. Therefore, this per month cost is $C(E[S],1/\beta,I)$. Comparing this cost function with the cost function of the *M/M/I/U* inspection regime, we see that there is one key difference, and this difference arises because the expressions for the $E[S]$ argument in these two functions are dissimilar (see equations (5) and (10)). As in section 3.2, the aggregate benefit and cost functions once again depend on a system aggregate, i.e., on $E[S]$ and on the mean inspection time $1/\beta$. Further, the only difference between the total benefit and cost functions lies in the third argument in each of these two functions. Specifically, this third argument is t_k (total tonnage being loaded/unloaded from the *kth* ship) in the aggregate benefit function and it is I (total number of inspectors) in the aggregate cost function. Finally, for the same reasons as those given in section 3.2, we have assumed that the aggregate benefit and cost functions depend not on the actual number of ships S but instead on the average number of ships $E[S]$.

Keeping this discussion in mind, we can now state our publicly owned port manager's maximization problem. This manager chooses the number of inspectors I and the inspection rate β to solve (also see footnote 10)

$$\max_{\{I,\beta\}} \alpha(1-P_I) \sum_{k=1}^{k=\alpha(1-P_I)} B_k(E[S],1/\beta,t_k)\text{-}C(E[S],1/\beta,I. \tag{11}$$

The reader should note two things about this maximization problem. First, in the inspection regime of this section we have U=I. However, unlike the optimization problem analyzed in section 3.2, we now suppose that port capacity or U is *endogenous* in the long run. This is why it makes sense to differentiate the net benefit function in equation (11) with respect to the choice variable I. Second, $E[S]$ in equation (11) is now given not by equation (5) but instead by equation (10). Suppose that the solution to problem (11) yields an interior maximum. Then, excluding the complementary slackness condition, the Kuhn-Tucker conditions for a maximum are

$$\alpha(1\text{-}P_I)\sum_{\forall k}\{\frac{\partial B_k(\bullet)}{\partial E[S]}\frac{\partial E[S]}{\partial I}\} - [\alpha^2(1\text{-}P_I)B_{\alpha(1-P_I)}(\bullet)+\sum_{\forall k}B_k(\bullet)]\frac{\partial P_I}{\partial I} =$$

$$\frac{\partial C(\bullet)}{\partial E[S]}\frac{\partial E[S]}{\partial I}+\frac{\partial C(\bullet)}{\partial I} \qquad (12)$$

and

$$\alpha(1\text{-}P_I)\sum_{\forall k}\{\frac{\partial B_k(\bullet)}{\partial E[S]}\frac{\partial E[S]}{\partial \beta} - \frac{\partial B_k(\bullet)}{\partial(1/\beta)}\frac{1}{\beta^2}\} - [\alpha^2(1\text{-}P_I)B_{\alpha(1-P_I)}(\bullet)+$$

$$\alpha\sum_{\forall k}B_k(\bullet)]\frac{\partial P_I}{\partial \beta} = \qquad (13)$$

$$\frac{\partial C(\bullet)}{\partial E[S]}\frac{\partial E[S]}{\partial \beta} - \frac{\partial C(\bullet)}{\partial(1/\beta)}\frac{1}{\beta^2}.$$

The optimal number of inspectors I^* and the optimal inspection rate β^* solve equations (12) and (13) respectively. Jointly, these two equations demonstrate the essential tradeoff facing our port manager. We now study each of these first order necessary conditions in greater detail. Equation (12) tells us that in choosing the number of inspectors optimally, the port manager will equate the marginal benefit from economic activities (the LHS) with the marginal cost of precluding biological invasions (the RHS).

Studying the LHS of equation (12) in greater detail, we see that the marginal benefit from economic activities is, in fact, the weighted difference of two terms. The first term $\alpha(1-P_I)\sum_{\forall k}\{\partial B_k(\bullet)/\partial E[S]\}\{\partial E[S]/\partial I\}$ captures the *secondary* impact that the optimal number of inspectors has on the marginal benefit through the *E*[*S*] variable. The second term $[\alpha^2(1-P_I)B_{\alpha(1-P_I)}(\bullet)+\sum_{\forall k}B_k(\bullet)]\{\partial P_I/\partial I\}$ captures the *primary* effect that the optimal number of inspectors has on the marginal benefit through the steady state probability P_I.

Moving to the RHS of equation (12), we see that the marginal cost of averting biological invasions is the *sum* of two terms. The first term $\{\partial C(\bullet)/\partial E[S]\}\{\partial E[S]/\partial I\}$ captures the *secondary* effect that the optimal number of inspectors has on the marginal cost through the *E*[*S*] variable. The second term $\{\partial C(\bullet)/\partial I\}$ accounts for the *primary* impact that the optimal number of inspectors has on the marginal cost of precluding biological invasions. Generally speaking, equation (12) cannot be solved analytically. Therefore, it will typically be necessary to use numerical methods to calculate the optimal number of inspectors I^*.

Equation (13) gives us the optimal inspection rate β^*. As described previously in section 3.2, our port manager selects the inspection rate so as to equate the marginal benefit from inspection at the optimal rate, with the pertinent marginal cost. In particular, the marginal benefit (the LHS of equation (13)) is the weighted difference of two terms. The first term captures the *secondary* impact that the optimal inspection rate β^* has on the marginal benefit through the $E[S]$ variable and the mean inspection time $1/\beta$. The second term on the LHS of equation (13) captures the *primary* effect that the optimal inspection rate β^* has on the marginal benefit through the steady state probability P_I.

Continuing on to the RHS of equation (13), we see that the marginal cost of precluding biological invasions is the *difference* of two terms. The first term $\{\partial C(\bullet)/\partial E[S]\}\{\partial E[S]/\partial \beta\}$ captures the impact that the optimal inspection rate β^* has on the marginal cost through the $E[S]$ variable. The second term $\{\partial C(\bullet)/\partial (1/\beta\}\{1/\beta^2\}$ accounts for the impact that the optimal inspection rate β^* has on the marginal cost of averting biological invasions through the mean inspection time $1/\beta$. Like equation (12), equation (13) too cannot, generally speaking, be solved analytically. Hence, it will be necessary to use numerical means to ascertain the optimal inspection rate β^*. We now compare and contrast the optimality conditions emanating from our analysis of the *M/M/I/U* and the *M/M/I/I* inspection regimes.

5. Inspection Regimes: *M/M/I/U* versus *M/M/I/I*

Let us first focus on equations (7) and (12). Comparing these two optimality conditions for the two inspection regimes that we are studying in this chapter, we see that there are three essential differences. First, by assumption, in the optimization problem studied in section 3.2, the port capacity U is exogenous in the long run whereas in the section 4.2 problem, this port capacity is, once again by assumption, endogenous. This difference has implications for the choice of the optimal number of inspectors I^*. Second, the effective arrival rate in equation (7) is $\alpha(1-P_U)$ and the effective arrival rate in equation (12) is $\alpha(1-P_I)$. Because these two effective arrival rates are dissimilar, the marginal benefit (the LHSs) in these two optimality conditions will generally be different. This means that the marginal cost (the RHSs) in these two optimality equations will also be different, and hence, the optimal number of inspectors in these two regulatory regimes can be expected to be distinct. Finally, observation of equations (5) and (10) tells us that the expected number of ships in the *M/M/I/U* inspection regime will generally be greater than the expected number of ships in the *M/M/I/I* inspection regime. This means that the volume of economic activity and hence the likelihood of a biological invasion will be greater in the *M/M/I/U* regime and lesser in the *M/M/I/I* regime.

We now turn to equations (8) and (13). Once again, there are three differences that are worth highlighting. First, equation (8) embodies the assumption that the port capacity is exogenous and equation (13) embodies the assumption that the same port capacity is endogenous. Clearly, this difference will affect the choice of the optimal inspection rate β^* in the two regulatory regimes being studied. Second, the effective arrival rate in equation (8) is $\alpha(1-P_U)$ and the effective arrival rate in equation (13) is $\alpha(1-P_I)$. Because these two rates are dissimilar, the marginal benefit (the LHSs) and the marginal cost (the RHSs) in these two optimality conditions will generally be different, and therefore, the optimal inspection rate β^* in these two inspection regimes can be expected to be distinct. Finally, observation of equations (8) and (13) tells us that the second term on the LHSs of these two equations is dissimilar. In particular, the partial derivative multiplying the term in boxed brackets on the LHS of equation (8) is $\partial P_U/\partial\beta$. In contrast, the corresponding partial derivative multiplying the term in boxed brackets in equation (13) is $\partial P_I/\partial\beta$.

6. Conclusions

In this chapter we developed what we believe is a new framework for studying the problem of preventing biological invasions caused by ships transporting internationally traded goods between countries and continents. This new framework allowed us to study the problem of preventing a biological invasion from a long run perspective. Specifically, we first characterized two regulatory regimes as two different kinds of queues. We then showed how a publicly owned port manager's decision problem can be posed and analyzed as an optimization problem using queuing theoretic techniques. Finally, we compared and contrasted the optimality conditions arising from our examination of the *M/M/I/U* and the *M/M/I/I* inspection regimes.

The analysis contained in this chapter can be extended in a number of different directions. In what follows, we suggest two possible extensions of this chapter's research. First, the reader will note that we analyzed Markovian inspection regimes, in this chapter. Therefore, it would be useful to investigate the properties of more general inspection regimes in which the arrival of ships and/or the service times of inspectors are characterized by general distribution functions. Second, on the numerical front, it would be useful to compare the approach of this chapter—in which the optimal number of inspectors choice problem is viewed as a continuous choice problem—with an alternate approach in which this choice problem is cast as an integer programming problem. Studies of international trade driven biological invasions that incorporate these aspects of the prevention problem into the analysis will provide additional insights into a phenomenon that has frequently proved to be very costly for the involved parties.

References

Barbier, E.B. 2001. A note on the economics of biological invasions, *Ecological Economics*, 39, 197-202.

Batabyal, A.A. 1996. The queuing theoretic approach to groundwater management, *Ecological Modelling*, 85, 219-227.

Cox, G.W. 1993. *Conservation Ecology*. W.C. Brown Publishers, Dubuque, IA.

de Wit, M.P., Crookes, D.J., and van Wilgen, B. W. 2001. Conflicts of interest in environmental management: Estimating the costs and benefits of a tree invasion, *Biological Invasions*, 3, 167-178.

Eiswerth, M.E., and Johnson, W.S. 2002. Managing nonindigenous invasive species: Insights from dynamic analysis, *Environmental and Resource Economics*, 23, 319-342.

Eiswerth, M.E., and van Kooten, G.C. 2002. Uncertainty, economics, and the spread of an invasive plant species, *American Journal of Agricultural Economics*, 84, 1317-1322.

Gross, D., and Harris, C.M. 1974. *Fundamentals of Queuing Theory*. John Wiley, New York, NY.

Heywood, V. Ed. 1995. *Global Biodiversity Assessment*. Cambridge University Press, Cambridge, UK.

Horan, R.D., Perrings, C., Lupi, F., and Bulte, E.H. 2002. Biological pollution prevention strategies under ignorance: The case of invasive species, *American Journal of Agricultural Economics*, 84, 1303-1310.

Office of Technology Assessment (OTA). 1993. *Harmful Non-Indigenous Species in the United States*. OTA-F-565, Washington, DC.

Olson, L.J., and Roy, S. 2002. The economics of controlling a stochastic biological invasion, *American Journal of Agricultural Economics*, 84, 1311-1316.

Parker, I.M., Simberloff, D., Lonsdale, W.M., Goodell, K., Wonham, M., Kareiva, P.M., Williamson, M.H., Von Holle, B., Moyle, P.B., Byers, J.E., and Goldwasser, L. 1999. Impact: Toward a framework for understanding the ecological effects of invaders, *Biological Invasions*, 1, 3-19.

Perrings, C., Williamson, M., and Dalmazzone, S. 2000. Introduction in C. Perrings, M. Williamson, and S. Dalmazzone, (Eds.), *The Economics of Biological Invasions*. Edward Elgar, Cheltenham, UK.

Pimentel, D., Lach, L., Zuniga, R., and Morrison, D. 2000. Environmental and economic costs of nonindigenous species in the United States, *BioScience*, 50, 53-65.

Ross, S.M. 2003. *Introduction to Probability Models*, 8th edition. Academic Press, San Diego, CA.

Savidge, J.A. 1987. Extinction of an island forest avifauna by an introduced snake, *Ecology*, 68, 660-668.

Vitousek, P.M., D'Antonio, C.M., Loope, L.L., and Westbrooks, R. 1996. Biological invasions as global environmental change, *American Scientist*, 84, 468-478.

Wolff, R.W. 1989. *Stochastic Modeling and the Theory of Queues*. Prentice-Hall, Englewood Cliffs, NJ.

Wolsey, L.A. 1998. *Integer Programming*. John Wiley, New York, NY.

CHAPTER 3

A RATIONALE FOR THE DIFFERENTIAL REGULATORY TREATMENT OF IMPORTS WHEN INVASIVE SPECIES ARE A POTENTIAL PROBLEM

Abstract. The maritime trade related risks from invasive species in a nation depend, *inter alia*, on the countries from which goods are being imported. In the United States, the USDA's APHIS routinely uses inspections to screen arriving ships at seaports for invasive species. Given this state of affairs, we first use queuing theory to construct a simple model of maritime trade and seaport inspections with one importing nation (Home) and two exporting nations. We then derive a ratio criterion that specifies a condition under which it makes sense for a seaport manager in Home to grant favorable regulatory treatment to the imports from one or the other exporting country. This *country of origin rule* depends on the mean time it takes to inspect ships from the two exporting countries and on the dollar value of the imports from these two countries.

1. Introduction

International trade in agricultural and non-agricultural products has increased greatly in the past few decades. This is the result of a whole host of factors, not the least of which is the phenomenon that is generally referred to today as globalization. Maritime trade is a large fraction of all trade and today ships are routinely used to transport a variety of internationally traded goods between countries. Although international trade in goods between nations is generally beneficial to these nations, as Heywood (1995), Parker *et al.* (1999), Batabyal (2004), Batabyal *et al.* (2005), Batabyal and Nijkamp (2005), and Batabyal and Lee (2006) have pointed out, today there is increasing concern that in addition to transporting goods between countries, by means of their container cargo and their ballast water, ships are also unwittingly transporting all kinds of invasive (also known as alien or non-native) plant and animal species from one geographical region to another.

An important method of unintentional marine non-native species introduction is by means of the dumping of ballast water. Cargo ships typically carry ballast water in order to enhance maneuverability and stability when they are not carrying full loads. When these ships come into port, this ballast water must be discharged before cargo can be loaded. It is estimated that over 4000 species of invertebrates, algae, and fish are being moved around the world in ship ballast tanks every day.[11] These alien species have often been very successful in invading their new habitats and the resulting biological invasions have proved to be very costly to the countries in which these new habitats are located. For the United States alone, the magnitude of these costs is astounding. For instance, according to the Office of Technology Assessment (OTA (1993)), the Russian wheat aphid caused an estimated $600 million worth of crop damage between 1987 and 1989. More generally, Pimentel *et al.* (2000) have estimated the total costs of all non-native species to be around $137 billion per year.

It is important to recognize that, in addition to economic costs, invasive species also cause significant ecological damage. As Vitousek *et al.* (1996) and de Wit *et al.* (2001) have noted, alien species can alter ecosystem processes, act as vectors of diseases, and diminish biological diversity. In this regard, the work of Cox (1993) tells us that out of 256 vertebrate extinctions with a known cause, 109 are the result of biological invasions. Even a single invasive species can cause enormous damage. Savidge (1987) tells us that following an invasion of Guam by

11In this regard, consider the case of one country, namely, Canada. It has been estimated that as much as 13 billion gallons or 50 million metric tonnes of overseas ballast water enters Canadian coastal ports, every year. A recent study by the Smithsonian Environmental Research Center (SERC) in Edgewater, Maryland calculated that a liter of ballast water typically contains several billion organisms similar to viruses and up to 800 million bacteria. For more details on these issues, go to http://www.fundyforum.com/profile_archives and to the SERC website, www.serc.si.edu

the brown tree snake, all twelve of this island's bird species became extinct.

The point of this discussion is clear. Invasive species can be and frequently have been a huge menace to society. Given this state of affairs, researchers have increasingly begun to focus their attention on the subject of invasive species management. Barbier (2001) shows that the economic effect of a biological invasion can be determined by analyzing the nature of the interaction between the alien and the native species. Eiswerth and Johnson (2002) analyze an optimal control model of the management of an alien species stock. They show that the optimal level of management effort is sensitive to ecological factors that are species and site specific and stochastic. Olson and Roy (2002) have used a model of a stochastic biological invasion to examine conditions under which it is optimal to eradicate the alien species and conditions under which it is not optimal to do so. Eiswerth and van Kooten (2002) have shown that, even when hard data about the spread of an invasive species are unavailable, it is possible to use information provided by experts to formulate a model in which it is optimal to not eradicate but instead control the spread of an invasive species. Finally, Batabyal and Beladi (2006) have provided a queuing theoretic perspective on the optimal ship inspection problem faced by a seaport manager.

The above studies have undeniably increased our understanding of regulatory issues in the context of invasive species management. Even so, a number of important research questions remain unanswered. Therefore, it is no surprise that the United States Department of Agriculture (USDA), under the aegis of the Program of Research on the Economics of Invasive Species Management (PRESIM), has now initiated a competitive award program to fund research on these salient and unanswered research questions. As the fiscal 2004 competitive award program description points out, *inter alia*, the USDA "seeks research that would improve understanding of how alternative rules and procedures for importing similar or identical products by different types of importers affect the costs and benefits of different exclusion and control strategies to reduce [invasive species] risks related to trade" (USDA (2004, p. 8)). The purpose of our chapter is to shed light on this question.

In particular, given that (i) the maritime trade related hazards from invasive species in a nation depend in part on the countries from which goods are being imported and (ii) the fact that at least in the United States, the USDA's Animal and Plant Health Inspection Service (APHIS) routinely uses inspections to screen arriving ships at seaports for invasive species, we first use queuing theory to construct a simple model of maritime trade and seaport inspections with one importing nation (Home) and two exporting nations. We then derive a ratio criterion that specifies a condition under which it makes sense for a seaport manager in Home to grant favorable regulatory treatment to the imports from one or the other exporting country. This *country of origin rule* depends on the mean time it takes to inspect ships from the two exporting countries and on the dollar value of the imports from these two countries. The reader should note that the research described in this chapter is entirely consistent with the agenda for invasive species research recently proposed by Batabyal (2004).

The rest of this chapter is organized as follows. Section 2 delineates the queuing theoretic framework in detail. Section 3 first provides evidence to support our claim that under some circumstances, a case can be made for providing favorable regulatory treatment to the imports from one country. Next, this section focuses on the nexus between the dollar value of imports and mean ship inspection times and shows that this nexus constitutes the basis for the differential regulatory treatment that we have just alluded to. Finally, section 4 concludes and discusses ways in which the research of this chapter might be extended.

2. The Theoretical Framework

2.1. A primer on queuing theory

Generally speaking, the purpose of queuing theory is to mathematically analyze waiting lines or queues (see Gross and Harris

(1974), Wolff (1989), and Ross (2002; 2003)). At a very elementary level, all queuing models have three features. Specifically, they can be described by (i) a stochastic arrival process, (ii) a random service time or times distribution function, and (iii) a deterministic number of servers. The arrival process is often but not always described by the Poisson process. When this is the case, the times between successive arrivals are exponentially distributed and, as is well known, the exponential distribution is memoryless or Markovian in nature. Therefore, the Poisson arrival process is commonly described by the letter *M*.

The service times are also random and thus these times can, in principle, be arbitrarily distributed. However, these services times are often modeled with the exponential distribution function which is memoryless. Hence, in this case, the letter *M* is also used to represent the service time distribution function. Finally, the deterministic number of servers is typically denoted by some positive integer.

So, for instance, the notation *M/M/*1 refers to a queuing model in which the arrival process is Poisson, the service time is exponentially distributed, and there is a single server. Similarly, the notation *M/G/*1 refers to a queuing model in which the arrival process is Poisson, the service times are generally distributed, and there is one server. In the rest of this chapter, we shall work with a variant of the *M/G/*1 queuing model to study the differential treatment of imports when invasive species are a potential problem.

2.2. Maritime trade and port inspections

We analyze a scenario in which there are two exporting countries and one importing country. The importing country is called Home and the subsequent analysis is conducted from the perspective of the manager (our server) of a representative seaport in Home. Home imports goods from two separate countries and these two exporting nations transport their goods via ships to the seaport under study in Home. Ships from the two exporting countries carry their cargo and ballast water into the seaport in Home. The Home seaport manager's problem is that the cargo and/or the ballast water in these ships may or may not contain potentially injurious alien animal and/or plant species. Therefore, this seaport manager's job is to ensure, to the extent possible, that all cargo arriving in ships from the two exporting nations in his seaport and the ballast water in these ships are inspected for the presence of possibly deleterious alien species.[12]

12To keep the subsequent mathematics relatively simple, we shall take an expansive view of the seaport manager's duties. In principle, it is possible to separate the task of inspecting cargo from that of inspecting ballast water. However, this separation would add algebraic clutter to the underlying analysis but would not alter the basic point we wish to make. Hence, in what follows, we shall stick with our expansive view of the seaport manager's duties. See chapter 13 for a model in which the above mentioned "separation of tasks" is explicitly modeled.

Upon arrival at our seaport, ships from each of the two exporting countries must first join a queue and then wait in this queue until it is their turn to be inspected and then cleared. We assume that ships from a particular exporting country are identical, but that ships from different countries are not. To model this feature, we shall say that the ships under study at our seaport are either from country A or from country B. Now, the arrival of ships from countries A and B coincides with the arrival of a whole host of (potentially damaging) alien species. It is reasonable to suppose that the arrival rate of these invasive species is proportional to the arrival rate of the aforementioned ships. Therefore, in what follows, we shall not model these invasive species directly. Instead, we shall focus on the ships that bring—by means of their cargo and their ballast water—these species to our seaport. Given this interpretation, the arrival process of the ships in our seaport constitutes the arrival process for the variant of the $M/G/1$ queuing model that we are using in this chapter.

Ships from countries A and B arrive at our seaport in accordance with independent Poisson processes with rates θ_A and θ_B. Notionally, when a country i ship arrives at our seaport for inspection and subsequent clearance, D_i, $i=A,B$, the dollar value of this ship's imported cargo per unit time is received and noted by the seaport manager. This seaport manager recognizes that the *risk* of a biological invasion in Home as a result of one or more species from either country A or B

invading Home is *different*. We model this aspect of the problem in two ways. First, we suppose that it takes a *random* amount of time to inspect the various country A and country B ships. To this end, let us denote the amount of time taken to inspect country A and country B ships by the random variables S_A and S_B Second, we suppose that S_A and S_B are *non-identical* random variables with *dissimilar* distribution functions.

Now, given D_A, D_B, S_A, and S_B, can we find a criterion that specifies a condition under which it makes sense for our seaport manager to grant favorable regulatory treatment to the imports from one or the other country? Put differently, we seek a *country of origin rule* that tells us which country—more precisely, which country's ships—ought to be *favored* by the seaport manager. This is the central question that we now answer. However, before we move to the specifics of our answer, let us first be clear about the sense in which we are using the word "favored." In this regard, without loss of generality, we shall say that country A ships are favored or given inspection priority over country B ships if inspection never commences on a country B ship when a country A ship is waiting in queue. However, if a country B ship is already being inspected and a country A ship arrives, then the country B ship continues to be inspected until this inspection process is completed. In other words, while country A ships are favored or given inspection priority, there is no preemption once inspection on a country *B* ship has commenced.[13]

We are now ready to answer the question we posed in the first sentence of the previous paragraph. However, before we provide this answer, let us be clear on three points. First, we are interested in deriving a *country of origin rule* that provides a rationale for providing favorable regulatory treatment to one or the other country in the possible presence of alien species. We are *not* interested in computing the expected social cost of ships waiting in the seaport under study. Second, let $p=\theta_A/(\theta_A+\theta_B)$ denote the proportion of all ships that are from country *A* and let $(1-p)=\theta_B/(\theta_A+\theta_B)$ denote the proportion of all ships that are from country *B*. Also, let T_i^N, $i=A,B$, be the mean amount of time a country *i* ship spends waiting in queue to be inspected and then cleared. Then, expressions like $p\,D_A T_A^N+(1-p)\,D_B T_B^N$ cannot be interpreted as "expected social cost" expressions because the D_i, $i=A,B$ are *not* cost terms but, instead, the dollar value of the imported cargo per unit time from each of the two countries that is received and noted by the seaport manager in Home. Finally, although it is possible to answer the question posed in

13This kind of a queuing model is sometimes called a non-preemptive priority queuing model. For textbook treatments of this and related models, see Ross (2002; 2003). In this connection, the reader should not interpret the seaport manager's inspection duties literally. In other words, this manager does not inspect every ship himself. Instead, we suppose that he has available to him a team of inspectors. In this chapter, we shall not concern ourselves with the details of this team. Readers wishing to learn more about this team issue should consult Batabyal (2004) and Batabyal and Beladi (2006).

the first sentence of the previous paragraph in different ways, in our opinion, the following four step procedure is the most efficient and illuminating. Therefore, in the next section, this is the approach that we follow.

3. The Differential Regulatory Treatment of Imports

3.1. Which country to favor?

First, we mathematically characterize the expected amount of time a ship spends waiting in queue in our queuing model with favoritism. Let us denote this expected wait time by T^N. In addition, $p = \theta_A/(\theta_A+\theta_B)$ and $(1\text{-}p)=\theta_B/(\theta_A+\theta_B)$ are as explained in the last paragraph of section 2.2. Now, recall that the seaport manager's inspection times for the ships from the two exporting countries are random variables denoted by S_A and S_B. As such, let $E[S_i],=1/t_i$, $i=A,B$, denote the expected inspection times for ships from these two countries and let $v_A=\theta_A E[S_A]$ and let $v_B=\theta_B E[S_B]$.

The reader should note that we are *not* modeling the inspection aspect of the problem with a cost function. Instead, we are modeling this aspect of the problem with two random variables—to account for the differential country risk of invasion—that account for the stochastic amount of time it takes to inspect the ballast water and/or the cargo of the ships from the two countries *A* and *B*. It is possible to model these two probabilistic inspection times with two different exponentially distributed random variables. In this case, if the S_i, i=*A*,*B*, are distributed exponentially then $E[S_i]$,=$1/t_i$, i=*A*,*B*, would be the means of these two exponentially distributed random variables and the t_i, i=*A*,*B*, would be the rates or the parameters of the two exponential distributions. If the exports of one country, say country *A* are perishable, then we can model this in one of two possible ways. First, we can posit that $E[S_A]$ =$(1/t_A)<(1/t_B)$=$E[S_B] \rightarrow t_A > t_B$. If the two inspection times are exponentially distributed then this means that the rate or the parameter of the exponentially distributed inspection time for country *A* ships (t_A) is greater than the corresponding rate or parameter for country *B* ships (t_B). Second, we can let S_A be deterministic with the additional proviso that $S_A < E[S_B]$.[14]

Returning to our queuing model, we see that there are ships from two different countries of origin to contend with. Therefore, to calculate T^N we must average over ships from both the exporting countries. This tells us that

14The primary objective of this chapter is to derive a country of origin rule that provides a rationale for the differential regulatory treatment of imports when alien species are a potential problem. This rule—see Proposition 1 below—does not depend on the demand for the exports of countries A and B in Home and hence we do not model demand explicitly.

$$T^N = pT_A^N + (1-p)T_B^N, \tag{1}$$

where T_i^N, i=A,B, is as explained in the last paragraph of section 2.2. Now, from equations (3) and (5) in Batabyal and Yoo (2007), it follows that

$$T_A^N = \frac{\theta_A E\left[S_A^2\right]+\theta_B E\left[S_B^2\right]}{2(1-v_A)} \text{ and } T_B^N = \frac{\theta_A E\left[S_A^2\right]+\theta_B E\left[S_B^2\right]}{2(1-v_A)(1-v_A-v_B)}. \tag{2}$$

Using equations (1) and (2) together, we reason that

$$T^N = \frac{\{\theta_A E\left[S_A^2\right]+\theta_B E\left[S_B^2\right]\}\{p(1-v_A-v_B)+(1-p)\}}{2(1-v_A-v_B)(1-v_A)} \tag{3}$$

We now have a mathematical expression for the expected amount of time a ship spends waiting in queue. This also completes the first step in the four step procedure that we alluded to in the last paragraph of section 2.2.

Our next task is to ascertain the mean amount of time spent by a ship waiting in queue when there is *no* favoritism demonstrated by the seaport manager. Although one can model the absence of favoritism in several ways, the simplest way is to suppose that our seaport manager inspects and clears ships on a first-come-first-served or FIFO basis. When ship inspections are carried out on a FIFO basis, the queuing model with favoritism that we have been studying thus far reduces to the well known *M/G/*1 queuing model in which the arrival rate of ships is $\theta = \theta_A + \theta_B$. Therefore, using equation 8.36 in Ross (2002, p. 256) we can tell that the average amount of time spent by ships waiting in queue, T^N_{FIFO}, is

$$T^N_{FIFO} = \frac{\{\theta_A E[S_A^2] + \theta_B E[S_B^2]\}}{2(1 - v_A - v_B)} \bullet \frac{(1 - v_A)}{(1 - v_A)}. \tag{4}$$

With equation (4) in place, we have now completed the second step in the four step procedure that we referred to earlier.

Our third task is to determine when the expected wait of ships in queue with favoritism is less than the expected wait without any favoritism. Put differently, with regard to the wait time in queue, we would like to know when it makes sense for our seaport manager to favor ships from one or the other country. Mathematically, we want to ascertain when $T^N < T^N_{FIFO}$. To perform this task, let us use equations (3) and (4). Comparing these two equations, it is clear that

$$T^N < T^N_{FIFO} \leftrightarrow p(-v_A - v_B) \leq -v_A \leftrightarrow p v_B > (1\text{-}p$$

$$v_A \leftrightarrow \frac{\theta_A \theta_B E[S_B]}{\theta_A + \theta_B} > \frac{\theta_B \theta_A E[S_A]}{\theta_A + \theta_B} \leftrightarrow E[S_B] > E[S_A] \leftrightarrow t_A > t_B. \quad (5)$$

Equation (5) provides the answer to the question we posed at the beginning of the previous paragraph. Specifically, this equation tells us that the average wait of ships in the queuing model with favoritism is less than the average wait of ships in the queuing model without favoritism if and only if the average time it takes to inspect country *B* ships is greater than the average time it takes to inspect country *A* ships. In other words, from a waiting time perspective, granting favorable preferential treatment to *A* ships over *B* ships makes sense if and only if the mean inspection time for *B* ships is greater than the mean inspection time for *A* ships. We are now in a position to undertake the fourth and final step in the four step procedure that we have been following thus far.

Recall that when a country i ship arrives at our seaport for inspection and subsequent clearance, D_i, $i=A,B$, the dollar value of each ship's imported cargo per unit time is received and noted by the seaport manager. Equation (5) tells us that if $t_A > t_B$ then $T^N < T^N_{FIFO}$. In words, when $t_A > t_B$, it makes sense for the seaport manager to favor country *A* ships because this favoritism or differential regulatory treatment minimizes the average wait time of ships in queue. Now, without going through all the mathematical minutiae, note that the argument that led to the above stated result from equation (5) also applies when $D_A t_A > D_B t_B$. In other words, $D_A t_A > D_B t_B$ implies that $T^N < T^N_{FIFO}$. This gives us the answer to the "Which country to favor?" question and we now state this answer as

PROPOSITION 1. In a situation of favoritism or differential regulatory treatment, if $E[S_A]/D_A < E[S_B]/D_B$ then the seaport manager ought to favor country A ships over country B ships when inspecting ships for potentially damaging invasive species.

3.2. Why favor?

There are two aspects to the problem of regulating invasive species. First, there is the trade or economic aspect. This aspect derives from the fact that the arrival of invasive species in Home cannot be separated from the arrival of imports by means of ships. Therefore, from a narrow economic perspective, our seaport manager should be taking those steps that permit the unloading and loading of cargo and the subsequent transfer of cargo to inland destinations to take place as efficiently as possible. Looked at from this perspective, activities—such as laborious inspections—that impede the trade related flow of economic activities are undesirable. In contrast, the second aspect of the regulatory problem is the biological aspect. In this regard, it is well known (see section 1) that invasive species that invade a new habitat successfully cause tremendous damage. In addition, as we have already noted, the arrival of these alien species cannot be separated from the arrival of imports. These two features together call for stringent inspections to eliminate or substantially reduce the likelihood of one or more invasions by alien species.

The discussion in the previous paragraph tells us that the economic and the biological aspects of the regulatory problem pull in *opposite* directions. The economic aspect suggests non-intrusive inspections but the biological aspect calls for detailed inspections. In addition, a perusal of USDA (2000) tells us that the risks from invasive species often do vary by country. Given these pieces of information and keeping the efficiency of economic activities in mind, how should a port manager discriminate between imports from different nations? We focused on two economic criteria (D_A, D_B) and two proxies for the risk of invasion (S_A, S_B) to answer this question by way of Proposition 1. This Proposition tells us that as far as inspections are concerned, country *A* ought to be favored over country *B* because the ratio of the risk of invasion proxy to the economic criterion ($E[S_A]/D_A$) for this country is lower than the corresponding ratio $E[S_B]/D_B$) for country *B*. In a two exporting countries and one importing country setting in which invasive species are a potential problem, Proposition 1 provides our seaport manager with a *country of origin rule* for the differential regulatory treatment of imports.

It is not difficult to show that if the ratio of the average inspection time to the dollar value of imports received from country *A* ships is greater than the corresponding ratio for country *B* ships then country *B* ships ought to be favored by our seaport manager in the inspection process. From a mathematical point of view, if $E[S_A]/D_A >$ $E[S_B]/D_B$ then, when conducting inspections, our seaport manager should discriminate in favor of country *B* ships over country *A* ships. This brings to an end our discussion of the "Why to favor?" question.

Before concluding this section, we now briefly comment on the situation in which Home anticipates increasing the volume of trade with one of the two exporters, say country *A* so that the value of the arrival rate θ_A is not constant but changing over time. One way to account for this feature would be to model the arrival process for country *A* ships not with a stationary or homogenous Poisson process but with a non-stationary or non-homogenous Poisson process with intensity function, say, $\theta_A(t)$, $t \geq 0$. By modeling the arrival process of ships from country *A* in this way, we would be explicitly accounting for the fact that exports from country A are more likely to arrive at certain times than at other times.

4. Conclusions

In this chapter, we used a queuing theoretic approach to analyze a question deemed to be significant by the USDA's PRESIM. Specifically, we derived a criterion that specifies a condition under which it makes sense for the Home seaport manager to grant favorable regulatory treatment to the imports from country A or B. To operationalize this rule, the seaport manager will need to compare the ratios given in Proposition 1. This comparison will enable the manager to show that a theoretical case can be made for favoring the ships from one country over the ships from the second country. Our analysis shows that the nature of this differential regulatory treatment of imports depends not only on specific economic criteria (D_A, D_B) but also on the proxies for the risk of a biological invasion (S_A, S_B).

The analysis in this chapter can be extended in a number of different directions. In what follows, we suggest three possible extensions. First, we analyzed a model in which there are two exporting countries. Therefore, one way to extend the analysis in this chapter would be to study a model in which there are $n \in \mathbb{N}$ exporting countries. It should be possible to use the approach of Batabyal and Yoo (2006) in another context to undertake this more general exercise. Second, it would be interesting to build on the extant Batabyal and Beladi (2006) analysis and study the problem of this chapter with a multi-server or multi-manager queuing model. Finally, in any inspection process, it is always possible that the captains of ships from some exporting countries will attempt to bribe the Home seaport manager to sanction negligent inspections. In these sorts of situations, game-theoretic modeling approaches are likely to be useful. Studies that analyze these aspects of the problem will strengthen our understanding of the nexuses between international maritime trade and invasive species management.

References

Barbier, E.B. 2001. A note on the economics of biological invasions, *Ecological Economics*, 39, 197-202.

Batabyal, A.A. 2004. A research agenda for the study of the regulation of invasive species introduced unintentionally via maritime trade, *Journal of Economic Research*, 9, 191-216.

Batabyal, A.A., and Beladi, H. 2006. International trade and biological invasions: A queuing theoretic analysis of the prevention problem, *European Journal of Operational Research*, 170, 758-770, 2006.

Batabyal, A.A., Beladi, H., and Koo, W.W. 2005. Maritime trade, biological invasions, and the properties of alternate inspection regimes, *Stochastic Environmental Research and Risk Assessment*, 19, 184-190.

Batabyal, A.A., and Lee, D.M. 2006. The infinitesimal, the deterministic, and the probabilistic: Alternate container inspection policies in invasive species management, *Biological Invasions*, 8, 1663-1671.

Batabyal, A.A., and Nijkamp, P. 2005. On container versus time based inspection policies in invasive species management, *Stochastic Environmental Research and Risk Assessment*, 19, 340-347.

Batabyal, A.A., and Yoo, S.J. 2006. A complete characterization of mean wait times for citizens in the nonpreemptive corruption regime, *Applied Economics Letters*,13, 759-762.

Batabyal, A.A., and Yoo, S.J. 2007. Corruption, bribery, and wait times in the public allocation of goods in developing countries, *Review of Development Economics*, 11, 507-517.

Cox, G.W. 1993. *Conservation Ecology*. W.C. Brown Publishers, Dubuque, IA.

de Wit, M.P., Crookes, D.J., and van Wilgen, B. W. 2001. Conflicts of interest in environmental management: Estimating the costs and benefits of a tree invasion, *Biological Invasions*, 3, 167-178.

Eiswerth, M.E., and Johnson, W.S. 2002. Managing nonindigenous invasive species: Insights from dynamic analysis, *Environmental and Resource Economics*, 23, 319-342.

Eiswerth, M.E., and van Kooten, G.C. 2002. Uncertainty, economics, and the spread of an invasive plant species, *American Journal of Agricultural Economics*, 84, 1317-1322.

Gross, D., and Harris, C.M. 1974. *Fundamentals of Queuing Theory*. John Wiley, New York, NY.

Heywood, V. Ed. 1995. *Global Biodiversity Assessment*. Cambridge University Press, Cambridge, UK.

Office of Technology Assessment (OTA). 1993. *Harmful Non-Indigenous Species in the United States*. OTA-F-565, Washington, DC.

Olson, L.J., and Roy, S. 2002. The economics of controlling a stochastic biological invasion, *American Journal of Agricultural Economics*, 84, 1311-1316.

Parker, I.M., Simberloff, D., Lonsdale, W.M., Goodell, K., Wonham, M., Kareiva, P.M., Williamson, M.H., Von Holle, B., Moyle, P.B., Byers, J.E., and Goldwasser, L. 1999. Impact: Toward a framework for understanding the ecological effects of invaders, *Biological Invasions*, 1, 3-19.

Pimentel, D., Lach, L., Zuniga, R., and Morrison, D. 2000. Environmental and economic costs of nonindigenous species in the United States, *BioScience*, 50, 53-65.

Ross, S.M. 2002. *Probability Models for Computer Science*. Academic Press, San Diego, CA.

Ross, S.M. 2003. *Introduction to Probability Models*, 8th edition. Academic Press, San Diego, CA.

Savidge, J.A. 1987. Extinction of an island forest avifauna by an introduced snake, *Ecology*, 68, 660-668.

United States Department of Agriculture (USDA). 2000. *Pest Risk Assessment for Importation of Solid Wood Packing Materials into the United States*. http://www.aphis.usda.gov/ppq/pra/swpm/complete.pdf

United States Department of Agriculture (USDA). 2004. *Program of Research on the Economics of Invasive Species Management Competitive Award Program: Description and Application Process*. Washington, DC.

Vitousek, P.M., D'Antonio, C.M., Loope, L.L., and Westbrooks, R. 1996. Biological invasions as global environmental change, *American Scientist*, 84, 468-478.

Wolff, R.W. 1989. *Stochastic Modeling and the Theory of Queues*. Prentice-Hall, Englewood Cliffs, NJ

CHAPTER 4

A DYNAMIC BERTRAND GAME MODEL OF TRADE, THREATS, AND EXPORT CERTIFICATION IN ALIEN SPECIES MANAGEMENT - WITH H. BELADI

Abstract. We analyze a dynamic Bertrand game model of trade between Foreign and Home in which the Foreign firm's export good is shipped to Home in containers carried by ships. This shipment to Home may also result in the unwitting introduction of alien species from Foreign into Home. By undertaking prophylactic measures in Foreign, the Foreign firm can *certify* that the container cargo it is sending to Home contains no alien species. However, because certification is *costly, ceteris paribus*, the Foreign firm has little or no incentive to certify its exports. Therefore, we investigate whether it is possible for a regulatory authority in Home to *threaten* to levy a duty on the Foreign firm's exports and thereby get this firm to certify its exports. Our analysis shows that when the elasticities of import and domestic demand and the profit function of the Foreign firm satisfy certain restrictions, it is indeed possible for the Home regulatory authority to obtain export certification with the threat of a duty on the Foreign firm's exports.

1. Introduction

It is now common knowledge that maritime trade in goods comprises a significant proportion of the world's total international trade in goods. Ships are the principal vehicle in maritime trade and hence they are customarily used to transport goods of all kinds, typically in containers, from one nation to another. Now, international trade theorists have demonstrated that there are benefits to the nations involved in such voluntary trade. Even so, in recent times, environmental and resource economists have claimed that these gains are likely to be smaller than what most researchers have hitherto believed. Why? As Perrings *et al.* (2000), Batabyal (2004), and Batabyal and Beladi (2006) have noted, this is because, in addition to transporting goods between nations, ships have also managed to carry an assortment of injurious alien animal and plant species (also known as exotic, invasive, and non-native species) from one part of the world to another.

The fact that exotic species have been and continue to be introduced into one region of the world from another is not new. What is new is the recognition that such introductions, particularly the unintentional ones, have often resulted in great losses to the affected nations. In this regard, consider the case of the United States. A report by the Office of Technology Assessment (OTA (1993)) noted that the annual monetary damage ensuing from biological invasions is between

$4.7 and $6.5 billion. More recent research by Pimentel *et al.* (2000) has concluded that the aggregate annual monetary damage from non-native species is, in fact, over $100 billion.

As far as unwitting introductions—the main focus of this chapter—are concerned, there are two essential ways in which alien species have been transported from one region of the world to another. First, many exotic species have been inadvertently introduced into a nation by ships disposing of their ballast water. Cargo ships routinely carry ballast water in order to enhance vessel stability when they are not carrying full loads. When these ships come into a seaport, this ballast water must be discarded, before cargo can be loaded. This manner of species introductions is salient and the problem of managing alien species that have been introduced into a particular nation by means of the discharge of ballast water has now received some attention in the economics literature.[15]

The second way in which alien species have been introduced into a particular nation is by means of contaminated goods—agricultural goods, fish, and fishery products readily come to mind—that are frequently transported in containers. In this regard, the reader should

15See Batabyal and Beladi (2006), Horan and Lupi (2005), Nunes and van den Bergh (2004), and Yang and Perakis (2004) for a more detailed corroboration of this claim.

note that alien species can remain hidden in containers for extended periods of time. In addition, material such as wood—that is often used to pack the cargo in the containers—may itself contain alien species. In fact, a joint report from the United States Department of Agriculture (USDA), the Animal and Plant Health Inspection Service (APHIS), and the United States Forest Service (USFS) has pointed out that nearly 51.8% of maritime shipments contain solid wood packing materials and that infection rates for solid wood packing materials are substantial (USDA, APHIS, and USFS (2000, p. 25)). For example, inspections of wooden spools from China revealed infection rates between 22% and 24% and inspections of braces for granite blocks imported into Canada were found to hold live insects 32% of the time (USDA, APHIS, and USFS (2000, pp. 27-28)).

Economists and ecologists are both very interested in effectively managing invasive species. As noted previously, this is because from an economic perspective, biological invasions can and often have resulted in large losses. In addition to these economic costs, the work of Vitousek *et al.* (1996), Costello and McAusland (2003), and others reminds us that alien species can change ecosystem processes, act as vectors of diseases, and decrease biological diversity. In this regard, Cox (1993) has noted that out of 256 vertebrate extinctions with a known cause, 109 are the result of biological invasions. The discussion in this paragraph clearly tells us that alien species have been and continue to be a great threat to society.

Despite the known nexus between goods trade between nations and the damage from alien species, with the exception of Jenkins (1996), ecologists in general have paid little attention to the role of trade policy in reducing the damage from alien species introductions. Jenkins (1996) has claimed that it may be necessary to use trade policy—such as trade bans and restrictions—to protect biological diversity. Very recently, a small number of researchers have begun to analyze issues at the interface of international trade and alien species management. Barbier and Shogren (2004) have studied a growth model in which a biological invasion occurs as a spillover effect from the importation of capital goods. They show that when a biological invasion decreases the productivity of all firms in the economy, the government ought to impose an output tax to equate the private and the social desires for consumption growth and capital accumulation. Costello and McAusland (2003) show that a tariff can either decrease or increase the damage from alien species. McAusland and Costello (2004) show that although it is always optimal to use tariffs to manage the damage from alien species, the same cannot be said about inspections. In particular, in their model, there are several instances in which it is optimal to not inspect imported goods at all. Finally, focusing on a trade model with four market structures, Batabyal and Beladi (2009) have identified conditions under which it makes sense to use tariffs to regulate alien species and conditions under which it does not.

The papers cited in the previous paragraph have, without a doubt, advanced our understanding of the effects of trade policy on the losses from alien species. Even so, an important, but unfortunately little studied, issue remains, and this is the issue of *export certification*. The basic issue here is this: By undertaking measures such as inspections in a foreign country, an exporting firm in the foreign country can *certify* that the container cargo it is shipping to the home country contains no alien species. However, because certification is *costly*, the foreign firm in general has little or no incentive to certify its exports. Therefore, in this chapter, we investigate whether it is possible for an apposite regulatory authority—who we shall refer to as an alien species manager—in the home country to threaten to levy a duty on the foreign firm's exports and thereby get this firm to certify its exports.

The rest of this chapter is organized as follows. Section 2.1 discusses some preliminary issues concerning export certification. Section 2.2 presents the dynamic Bertrand game model of trade between two countries, namely, Foreign and Home. Section 2.3 provides a detailed analysis of the conditions under which it is possible for the alien species manager in Home to obtain export certification, with the threat of a duty on the Foreign firm's exports. Section 3 concludes and offers suggestions for future research on the subject of this chapter.

2. The Theoretical Framework

2.1. Preliminaries

The basic idea behind export certifications of any kind is that the exporting firm will take certain actions, such as inspections in the exporting *country of origin*, i.e., in Foreign, that will provide the importing nation with a reasonable assurance about the quality of the underlying exports. As an example, consider the export of fish and fishery products by an Indian firm to the European Union that is discussed in Sawhney (2005). In this case, under certain conditions, the Export Inspection Council (EIC) of India will certify that the export shipments of the Indian firm satisfy certain international standards, such as the food safety management systems based certification.

Now, in the aftermath of the dreadful events of 11 September 2001, a program very much like the export certification of goods has been instituted by the United States Customs and Border Protection (CBP), and this program is called the Container Security Initiative (CSI).[16] Although the CSI is primarily a national security initiative, like

16To learn more about the CSI, go to www.CBP.gov and download the web documents "CSI in Brief" and the CSI "Fact Sheet."

export certification in the context of agricultural and food trade, one of the key objectives of the CSI is to pre-screen and evaluate containers carrying cargo, at the point of origin, before they are shipped to the United States.

From the discussion in the previous two paragraphs, the reader will note that export certification programs exist not only in the context of agricultural goods trade but also in the context of goods trade more generally, when terrorism is an issue. In this chapter, we are interested in export certification programs in the context of alien species management. Although the focus here, once again, is on container cargo, the main concern now is *not* to ensure a certain level of quality or to minimize the risk of terrorism. Instead, the primary concern now is to ensure that the container cargo has been adequately screened in the exporting country of origin so that alien species are not transported from the various countries of origin (the exporting nations) to the importing nations.

Although export certification in the context of alien species management has been previously discussed, to the best of our knowledge, there are no theoretical analyses of such programs. In this regard, the reader should understand that export certification, although beneficial for importing nations, is *costly* for exporting firms. Therefore, *ceteris paribus*, they have little or no incentive to undertake expensive procedures, such as inspections, and thereby obtain the requisite export certification. In contrast, an alien species manager in an importing nation will, for obvious reasons, want exporting firms to be able to certify their exports. This notwithstanding, an alien species manager knows that not only can he not compel exporting firms to certify their exports but that he also cannot be sure—for the above cost related reasons—whether individual claims of export certification from exporting firms are credible.

The above informational drawback notwithstanding, the alien species manager in an importing nation does know the *price* of the export good and, given what steps a particular exporting firm says it has taken to obtain export certification, this manager also has a pretty good idea about the cost of these steps.[17] Therefore, on the reasonable assumption that exporting firms are profit maximizers, it is fair to say that an export good that has undergone certification will be priced *higher* than the same good without certification. This *differential pricing* idea is what we shall exploit as we analyze whether it is possible for an alien species manager in an importing nation to threaten to levy a duty on a foreign firm's exports and thereby get this firm to certify its exports.

17This is true because an alien species manager can use currently available information about the export certification costs of related activities—such as ensuring the quality of a food shipment or identifying containers that pose a high risk for terrorism—to form his impressions about the cost of procuring certification. See Kelly (2003), Sawhney (2005) and the references cited in footnote 16 for additional details on this point.

2.2. The two period Bertrand game model

The world consists of two countries called Home and Foreign. The Foreign firm and the Home firm engage in Bertrand competition. This Foreign firm exports and Home imports a particular—agricultural or manufactured—good. If the domestic and the imported goods are perfect substitutes, then it is well known—see Varian (1992, p. 292)—that Bertrand competition leads to marginal cost pricing. To avoid this case and to keep matters more interesting, we assume that the domestic and the import goods are *imperfect* substitutes. Let the price of the imported good and the domestic good in Home be p and q, respectively. In the rest of this chapter, we shall adopt the convention of labeling functions pertaining to Foreign with stars (*).

As indicated in section 2.1, the process of obtaining export certification is costly for the exporting firm in Foreign. Therefore, *ceteris paribus*, it has little or no incentive to go through this costly process. However, the alien species manager in Home will certainly want to see certification of the Foreign firm's exports. This not only diminishes the likelihood of one or more biological invasions in Home but it also reduces the inspection costs that this manager in Home would otherwise have to incur at the relevant port of entry. So, to encourage the Foreign firm to certify its exports, the alien species manager interacts with the exporting firm in Foreign over two time periods as follows:

Given the period 1 import price p_1 chosen by the exporting firm from Foreign, the alien species manager in Home imposes an *ad valorem* duty in period 2 equaling $(1+\tau_2)= \hat{p}/p_1$ whenever the first period import price p_1 is less than the price $\hat{p}$. Here, the benchmark price $\hat{p}$ is our alien species manager's best estimate of what the price of the Foreign firm's export good would be if its exports had, in fact, been certified. The interesting thing about the alien species manager's behavior is that he does not impose the above specified *ad valorem* duty definitively. Instead, he *threatens* to impose this duty whenever $p_1 < \hat{p}$. Mathematically, this means that the duty is imposed with probability $\alpha >$ 0 and with complementary probability $(1-\alpha)$ there is no duty on the Foreign firm's export good. The reader should understand that, because all other reasons for wanting to raise the first period price p_1 are ruled out, getting this exporting firm to raise p_1 is equivalent to getting the firm to certify its exports. The basic question that we now need to answer is whether the above kind of threat will, in fact, succeed in getting the Foreign firm to undertake export certification. To reiterate, mathematically what we want to know is whether the Home alien species manager's threat will result in an *increase* in the Foreign exporting firm's first period price p_1.

2.3. Analysis

To answer the question we have just posed it will be necessary to analyze the Bertrand game between the Foreign exporting firm and the Home firm. To this end, let us now introduce some additional notation. Let $\pi_2^*(\tau_2)$ denote the Foreign exporting firm's second period profit when the alien species manager in Home imposes his duty. Also, let π_2^* (0) denote the maximized value of the Foreign exporter's second period profit when there is no duty imposed by the Home alien species manager. Now, before we proceed further with the Foreign exporter's first period profit maximization problem, it will be necessary to determine the sign of the derivative $\pi_2^{*\prime}(\tau_2)$. To this end, we now state

Lemma 1. The Foreign exporting firm's profit is a decreasing function of the duty or $\pi_2^{*\prime}(\tau_2)<0$.

Proof. We begin by writing the Foreign exporting firm's second period profit function. That function is

$$\pi_2^* = \frac{p_2 d^*(p_2, q_2, I)}{(1+\tau_2)} - C^*\{d^*(p_2, q_2, I_2)\}, \tag{1}$$

where $d^*(\bullet)$ is the demand function for the imported good in Home, I_2 is the income spent on the imported good and the domestic good in the second period in Home, and $C^*\{\bullet\}$ is the Foreign exporter's cost function. Now, to simplify the proof, we suppose that the demand functions for the imported good $\{d^*(\bullet)\}$ and the domestic good $\{d(\bullet)\}$ have unitary income elasticities. Then, as noted by Feenstra (2004, p. 229), the income spent on the two goods in the second period I_2 does *not* affect the price elasticities or the two firm reaction functions. Therefore, the positive domestic and foreign price elasticities $\phi = -\{\partial d(\bullet)/\partial q_2)(q_2/d(\bullet))\}$ and $\phi^* = -\{\partial d^*(\bullet)/\partial p_2)(p_2/d^*(\bullet))\}$ can be written as functions of the price ratio of p_2 and q_2 or $\phi(q_2/p_2)$ and $\phi^*(p_2/q_2)$.

With the above characterization of the two price elasticities in place, let us now differentiate equation (1) with respect to the duty τ_2. We get

$$\frac{d\pi^*}{d\tau_2} = \pi^*_{p_2}\frac{dp_2}{d\tau_2} + \pi^*_{q_2}\frac{dq_2}{d\tau_2} + \pi^*_{\tau_2} = \{\frac{p_2 d^*_{q_2}(\bullet)}{1+\tau_2} + C^{*'}(\bullet)d^*_{q_2}(\bullet)\}\frac{dq_2}{d\tau_2} - \frac{p_2 d^*(\bullet)}{(1+\tau_2)^2}, \quad (2)$$

where we have simplified the second equality in the above expression by using the first order condition $\pi^*_{p_2}$ =0. Now, after several steps of algebra, it can be shown that

$$\frac{d\pi^*}{d\tau_2}\frac{(1+\tau_2)}{d^*(\bullet)}=\frac{q_2 d^*_{q_2}(\bullet)}{d^*(\bullet)}\{\frac{p_2}{(1+\tau_2)}-C^{*'}(\bullet)\}\frac{dq_2}{d\tau_2}\frac{(1+\tau_2)}{q_2}-\frac{p_2}{(1+\tau_2)}. \quad (3)$$

Using the previously stated expressions for the two price elasticities ϕ and ϕ^*, the right-hand-side (RHS) of equation (3) can be simplified further. This simplification yields

$$\frac{d\pi^*}{d\tau_2}\frac{(1+\tau_2)}{d^*(\bullet)}=\{\frac{(\phi^*-1)}{\phi^*}\frac{dq_2}{d\tau_2}\frac{(1+\tau_2)}{q_2}-1\}\frac{p_2}{(1+\tau)}. \quad (4)$$

In writing equation (4), we have used two pieces of information. The first piece of information is that $d^*_{q_2}(\bullet)q_2/d^*(\bullet)=(\phi^*-1)$. The second piece of information is the first order necessary condition to the Foreign exporting firm's profit maximization problem (see equation (1)) $p_2\{1-(1/\phi^*)\}=(1+\tau_2)C^{*'}(\bullet)\Rightarrow\{p_2/(1+\tau_2)-C^{*'}(\bullet)\}=p_2/\{\phi^*(1+\tau_2)\}$. Now, from Feenstra (2004, p. 252) it follows that $(dq_2/d\tau_2)\{(1+\tau_2)/q_2\}<1$. Using this last inequality in equation (4) it is clear that the Foreign exporting firm's profit function $\pi_2^*(\bullet)$ is decreasing in the duty τ_2 or $\pi_2^{*'}(\tau_2)<0$.

With Lemma 1 in place, we can now state the Foreign exporting firm's first period profit maximization problem. This firm solves

$$\max_{\{p_1\}} \pi^*(p_1, q_1)+\delta[\alpha\pi_2^*(\tau_2)+(1-\alpha)\pi_2^*(0)], \tag{5}$$

where δ is the per period discount factor and the constraint is $(1+\tau_2)=\hat{p}/p_1$. In like manner, the Home firm chooses q_1 to maximize $\pi(p_1, q_1)$. The first order necessary conditions to the two profit maximization problems specified above are

$$\frac{\partial\pi^*(p_1,q_1)}{\partial p_1}=\delta\alpha\frac{d\pi_2^*(\tau_2)}{d\tau_2}\frac{\hat{p}}{p_1^2}<0 \text{ and } \frac{\partial\pi(p_1,q_1)}{\partial q_1}=0. \tag{6}$$

Inspecting equation (6) we see that the first order necessary condition for the Foreign firm has a negative term—$\pi_2^{*'}(\tau_2)$—on the right. Now, using the second order sufficiency condition $\pi^*_{p_1p_1}(\bullet)<0$ it is clear that this will *shift* the Foreign firm's reaction function to the *right* from R*R* to R'R' as

Figure 1

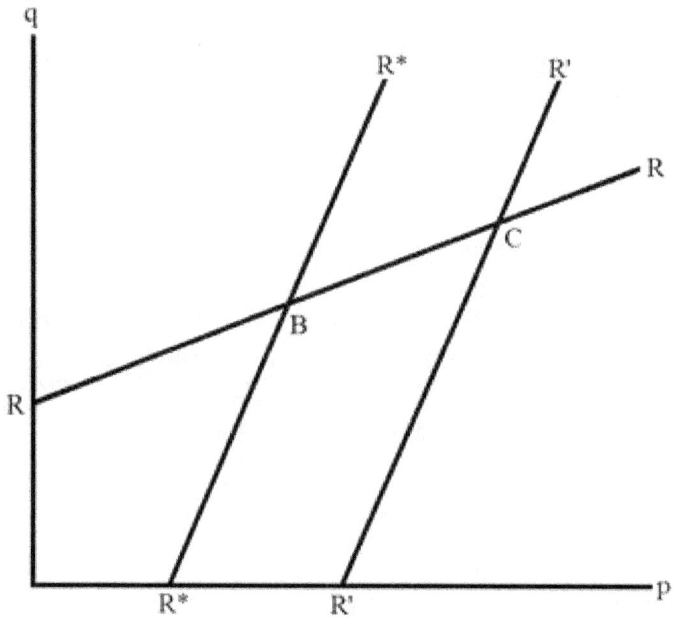

shown in Figure 1. This tells us that the first period import price does *go up* from point B in Figure 1—the no duty equilibrium of the Bertrand game between the two firms—to point C because of the credible threat of a duty by the alien species manager in Home. RR in Figure 1 is the Home firm's reaction function and the two prices, p and q are plotted on the horizontal and the vertical axes. We have now answered the central question of this chapter. Specifically, our analysis thus far tells us that if the positive domestic and foreign price elasticities ϕ and ϕ^* can be written as functions of the price ratio of p_2 and q_2 *and* if the second period profit function $\pi_2^*(\tau_2)$ of the Foreign firm is decreasing in the duty τ_2 then it is possible for the alien species manager in Home to raise the first period price of the Foreign firm's export good. As indicated in section 2.2, in this chapter, this is equivalent to obtaining export certification from the Foreign firm. Therefore, to sum up, if the above mentioned restrictions on the two price elasticities and the Foreign firm's profit function hold, then it is possible for the alien species manager in Home to obtain export certification by threatening to levy a duty on the Foreign export good if this good's first period price p_1 is less than the manager's benchmark price of $\hat{p}$.

3. Conclusions

In this chapter, we analyzed a dynamic Bertrand game model of trade between Foreign and Home in which the Foreign firm's export good is shipped to Home in containers carried by ships. This shipment of the export good to Home may also result in the unwitting introduction of alien species from Foreign into Home. By undertaking certain *costly* measures in Foreign, the Foreign firm can *certify* that the container cargo it is shipping to Home contains no alien species. In this setting, we investigated whether it is possible for an alien species manager in Home to threaten to levy a duty on the Foreign firm's exports and thereby get this firm to certify its exports. Our analysis showed that when the elasticities of import and domestic demand and the profit function of the Foreign firm satisfy certain restrictions, it is indeed possible for the alien species manager in Home to obtain export certification with the threat of a duty on the Foreign firm's exports.

The analysis in this chapter can be extended in a number of directions. In what follows, we suggest two possible extensions of this chapter's research. First, in our model the threat of a duty by the alien species manager in Home results in the Foreign exporting firm undergoing export certification and, equivalently, raising the price of its export good. In other words, this firm passes the cost of obtaining export certification to consumers in Home and, in the process, makes its own position on the international market less competitive. This sequence of events need not take place. The Foreign government may bear the cost of export certification and in this case a "low" price of the export good would not necessarily mean that the Foreign firm had not undergone export certification. Therefore, it would be useful to analyze a scenario in which the alien species manager in Home uses price *and* other information to determine whether or not to levy a duty on the Foreign firm's export good. Second, in the context of a potential two way flow of alien species from Foreign to Home and from Home to Foreign, it would be useful to analyze a scenario in which the two nations *jointly* determine protocols for managing invasive species. Studies of international trade and alien species management that incorporate these aspects of the problem into the analysis will provide additional insights into a management problem that has considerable economic and ecological ramifications.

References

Barbier, E.B., and Shogren, J.F. 2004. Growth with endogenous risk of biological invasion, *Economic Inquiry*, 42, 587-601.

Batabyal, A.A. 2004. A research agenda for the study of the regulation of invasive species introduced unintentionally via maritime trade, *Journal of Economic Research*, 9, 191-216.

Batabyal, A.A., and Beladi, H. 2006. International trade and biological invasions: A queuing theoretic analysis of the prevention problem, *European Journal of Operational Research*, 170, 758-770.

Batabyal, A.A., and Beladi, H. 2009. Trade, the damage from alien species, and the effects of protectionism under alternate market structures, *Journal of Economic Behavior and Organization*, 70, 389-401.

Costello, C., and McAusland, C. 2003. Protectionism, trade, and measures of damage from exotic species introductions, *American Journal of Agricultural Economics*, 85, 964-975.

Cox, G.W. 1993. *Conservation Ecology*. W.C. Brown Publishers, Dubuque, IA.

Feenstra, R.C. 2004. *Advanced International Trade*. Princeton University Press, Princeton, NJ.

Horan, R.D., and Lupi, F. 2005. Tradeable risk permits to prevent future introductions of invasive alien species into the great lakes, *Ecological Economics*, 52, 289-304.

Jenkins, P. 1996. Free trade and exotic species introductions, *Conservation Biology*, 10, 300-302.

Kelly, T. 2003. The WTO, the environment, and health and safety standards, *World Economy*, 26, 131-151.

McAusland, C., and Costello, C. 2004. Avoiding invasives: Trade-related policies for controlling unintentional exotic species introductions, *Journal of Environmental Economics and Management*, 48, 954-977.

Nunes, P.A.L.D., and van den Bergh, J.C.J.M. 2004. Can people value protection against invasive marine species? Evidence from a joint TC-CV survey in the Netherlands, *Environmental and Resource Economics*, 28, 517-532.

Office of Technology Assessment (OTA). 1993. *Harmful Non-Indigenous Species in the United States*. OTA-F-565, Washington, DC.

Perrings, C., Williamson, M., and Dalmazzone, S. 2000. Introduction in C. Perrings, M. Williamson, and S. Dalmazzone, (Eds.), *The Economics of Biological Invasions*. Edward Elgar, Cheltenham, UK.

Pimentel, D., Lach, L., Zuniga, R., and Morrison, D. 2000. Environmental and economic costs of nonindigenous species in the United States, *BioScience*, 50, 53-65.

Sawhney, A. 2005. Quality measures in food trade: The Indian experience, *World Economy*, 28, 329-348.

USDA, APHIS, and USFS. 2000. Pest Risk Assessment for Importation of Solid Wood Packing Materials into the United States. http://www.aphis.usda.gov/ppq/pra/swpm/complete.pdf

Varian, H.R. 1992. *Microeconomic Analysis*, 3rd edition. W.W. Norton and Company, New York, NY.

Vitousek, P.M., D'Antonio, C.M., Loope, L.L., and Westbrooks, R. 1996. Biological invasions as global environmental change, *American Scientist*, 84, 468-478.

Yang, Z., and Perakis, A.N. 2004. Multiattribute decision analysis of mandatory ballast water treatment measures in the US great lakes, *Transportation Research Part D*, 9, 81-86.

CHAPTER 5

TRADE, THE DAMAGE FROM ALIEN SPECIES, AND THE EFFECTS OF PROTECTIONISM UNDER ALTERNATE MARKET STRUCTURES - WITH H. BELADI

Abstract. We first construct three measures of the expected damage from the unintentional introduction of alien species into a country called Home. We then focus on four market structures. First, perfect competition prevails in both Home and Foreign, and Home is a small country. Second, the Home and the Foreign markets are both perfectly competitive, but Home is now a large country. Third, the exporter in Foreign is a monopolist and there are no import competing firms in Home. Finally, the Foreign exporter and the import competing firm in Home engage in Cournot competition. In all four scenarios, we analyze the impact of small and optimal Home tariffs on prices, exports, imports, the damage from alien species, and social welfare in Home. *Inter alia*, our analysis identifies conditions under which it makes sense to use trade policy (tariffs) to regulate invasive species and conditions under which it does not.

1. Introduction

The fact that alien species (also known as invasive or non-native species) have been and continue to be introduced into one part of the world from another is not new. What is new is the realization that such introductions, particularly the unintentional ones, have often been very costly for the concerned nations. In this regard, consider the case of the United States. A report by the Office of Technology Assessment (OTA 1993) declared that the annual monetary damage resulting from biological invasions is between \$4.7 and \$6.5 billion. More recent research by Pimentel *et al.* (2000) has concluded that the total annual monetary damage from invasive species is in fact over \$100 billion.

Researchers now recognize that maritime trade in goods comprises a sizeable proportion of the world's total international trade in goods. Ships are the primary vehicle in maritime trade, and consequently they are routinely used to carry goods of all kinds (often in containers) from one country to another. Now, international trade theorists have demonstrated that there are benefits to the nations involved in such voluntary trade. This notwithstanding, in recent times, natural resource and environmental economists have contended that these gains are likely to be smaller than what most researchers have believed thus far. Why? As Perrings *et al.* (2000), Costello and McAusland (2003), Batabyal (2004), Batabyal *et al.* (2005), and Margolis *et al.* (2005) have noted, this is because in addition to carrying goods between nations, ships have also managed to carry an assortment of deleterious non-native plant and animal species from one part of the world to another.

As far as unintentional introductions, the primary focus of this chapter, are concerned, there are two main ways in which alien species have been carried from one part of the world to another. First, many invasive species have been introduced into a country, often inadvertently, by ships discarding their ballast water. Cargo ships usually carry ballast water in order to increase vessel stability when they are not carrying full loads. When these ships come into a seaport, this ballast water must be jettisoned, before cargo can be loaded. This manner of species introductions is important, and the problem of managing alien species that have been introduced into a particular nation by means of the discharge of ballast water has now received some attention in the economics literature (see Nunes and Van den Bergh 2004, Yang and Perakis 2004, and Batabyal and Beladi 2006).

The second way in which alien species have been introduced into a particular country is by means of contaminated goods (agricultural goods readily come to mind) that may or may not be carried in containers. In this regard, the reader should note that invasive species can remain concealed in containers for long periods of time. In addition, material such as wood, which is often used to pack the cargo in the containers, may itself contain alien species. In fact, as pointed out by Costello and McAusland (2003), a joint report from the United States Department of Agriculture (USDA), the Animal and Plant Health Inspection Service (APHIS), and the United States Forest Service (USFS) has noted that nearly 51.8% of maritime shipments contain solid wood packing materials and that infection rates for solid wood packing materials are substantial (USDA, APHIS, and USFS 2000, p. 25). For example, inspections of wooden spools from China revealed infection rates between 22% and 24%, and inspections of braces for granite blocks imported into Canada were found to hold live insects 32% of the time (USDA, APHIS, and USFS, 2000, pp. 27-28).

Economists and ecologists are both very interested in managing invasive species. This is because (see the first paragraph of this section) biological invasions can and often have proven to be very costly from an economic standpoint. In addition to these economic costs, the work of Vitousek *et al.* (1996), Simberloff *et al.* (1997), Costello and McAusland (2003), and others, reminds us that alien species can alter ecosystem processes, act as vectors of diseases, and diminish biological diversity. In this regard, Cox (1993) has observed that out of 256 vertebrate extinctions with a known cause, 109 are the outcome of biological invasions. This discussion tells us that non-native species have been and continue to be a great menace to society.

It is only very recently that economists have begun to analyze questions pertaining to invasive species management. For instance, Eiswerth and Johnson (2002) have studied an intertemporal model of alien species stock management. They note that the optimal level of management effort is responsive to ecological factors that are not only species and site specific but also stochastic in nature. Second, Olson and Roy (2002) have used a stochastic framework to examine the circumstances under which it is optimal to wipe out an invasive species and the circumstances under which it is not optimal to do so. Third, Horan *et al.* (2002) have analyzed the properties of management approaches under full information and under uncertainty. Fourth, Batabyal *et al.* (2005) have observed that there is a tension between economic cost minimization and inspection stringency in invasive species management. Finally, Batabyal and Beladi (2006) have analyzed maximization problems stemming from the steady state analysis of two multi-person inspection regimes.

Despite the known connection between goods trade between countries and the damage from alien species, with the exception of Jenkins (1996), ecologists in general have paid *scant attention* to the role of trade policy in mitigating the damage from alien species introductions. Jenkins (1996) has contended that it may be necessary to use trade policy (bans and restrictions) to protect biological diversity. Very recently, a small number of papers have begun to analyze issues at the interface of international trade and invasive species management. Barbier and Shogren (2004) have analyzed a growth model in which a biological invasion occurs as a spillover effect from the importation of capital goods. They show that when a biological invasion diminishes the productivity of all firms in the economy, the government ought to impose an output tax to equate the private and the social desires for consumption growth and capital accumulation. Costello and McAusland (2003) and McAusland and Costello (2004) have studied the impact that tariffs have on the damage from invasive species introductions. Costello and McAusland (2003) show that a tariff can either decrease or increase the damage from invasive species. McAusland and Costello (2004) show that although it is always optimal to use tariffs to control the damage from alien species, the same cannot be said about inspections. In particular, in their model, there are several circumstances in which it is optimal not to inspect imported goods at all. Prestemon *et al.* (2006) study international trade in forest products and show that trade liberalization will have a negligible effect on US imports of Siberian logs and on the risk of a biological invasion. Finally, using an integrated model with an international trade component, Zhao *et al.* (2006) demonstrate the consumer and the producer responses to livestock disease outbreaks and the welfare effects of alternate invasive species management policies.

Although the papers cited in the previous paragraph have certainly advanced our understanding of the impacts of trade policy on the damage from invasive species, three outstanding questions concerning the desirability of using trade policy to manage invasive species remain. Therefore, the purpose of this chapter is to analyze these three questions in detail. First, unlike the extant literature, we use a two country model to study the efficacy of tariffs as an invasive species management tool under four *different market structures*.[18] Second, we focus not just on small tariffs but on *small and on optimal* tariffs. Finally, our emphasis is less on the impact that tariffs have on the damage from invasive species *per se* and more on the *impacts of tariffs on social welfare* when social welfare depends in part on the damage from alien species.[19]

The rest of this chapter is organized as follows. Section 2.1 briefly describes our two country model. Sections 2.2 to 2.4 construct three measures of the expected damage in Home from the introduction of alien species. Section 3.1 continues the description of our two country model. Then, this section derives a general expression for the change in

18For analyses of tariffs in other contexts, see Parai (1999), Biswas and Marjit (2007), and Vishwasrao *et al.* (2007).

19We are not suggesting that this chapter is the first to study the impacts of tariffs on social welfare when social welfare depends in part on the damage from alien species. This issue has also been looked at by McAusland and Costello (2004) and by Margolis *et al.* (2005). Instead, what we are suggesting is that, to the best of our knowledge, this chapter is the first to study the three questions stated earlier in this paragraph collectively.

Home social welfare as a function of a change in the Home tariff. Section 3.2 analyzes the impact of small and optimal tariffs imposed by Home on the damage from invasive species, on prices, on exports and imports, and on social welfare in Home for the case in which perfect competition prevails in both Home and Foreign and Home is a small country. Section 3.3 does the same for the case in which the Home and the Foreign markets are both perfectly competitive, but Home is now a large country. Section 3.4 conducts a similar analysis for the case in which the exporter in Foreign is a monopolist and there are no import competing firms in Home. Section 3.5 also conducts the same kind of analysis as the previous three sections, but now the Foreign exporter and the import competing firm in Home engage in Cournot competition. Section 3.6 first discusses the form of the dependence of all the tariff expressions on the expected total damage from alien species introductions in Home. Next, this section comments briefly on scenarios in which the above discussed form of dependence would be different. Section 4 concludes and offers suggestions for future research on the subject of this chapter.

2. Three Measures of Damage in Home from Stochastic Alien Species Introductions

2.1. Preliminaries

The world consists of two countries called Home and Foreign. Foreign exports and Home imports a specific good that could be either an agricultural good or a manufactured good. Over time, the import of this good also results in the *probabilistic* introduction of alien species from Foreign into Home. Initially, because of scientific uncertainty, citizens and the relevant authorities in Home do not realize that these unintentionally introduced alien species cause agricultural and/or ecological damage in Home. However, with the passage of time, scientific evidence implicating the alien species emerges, and then it becomes clear to the citizens and to the aforementioned authorities that the stochastic introductions of these alien species and the resulting monetary damage are *linked* to the import of the good in question from Foreign. With this realization come calls for the use of trade policies to restrict imports and thereby reduce the introduction of the deleterious alien species.

Given this temporal sequence of events, we now construct three measures of the expected damage in Home from the stochastic introduction of alien species. The measures in sections 2.2 and 2.3 are monetary measures of damage, and the measure in section 2.4 is a physical measure of damage. The damage measures in sections 2.2 and 2.4 are based on the work of Batabyal and Nijkamp (2007), and the damage measure in section 2.3 is based on the analysis in Batabyal and Beladi (2001). Why are we focusing on three measures of damage? This is because we would like to ascertain whether alternate specifications of the damage metric have a similar or a dissimilar qualitative impact on the various small and optimal tariff expressions that we derive in sections 3.2 to 3.5 below.

2.2. First Measure of Damage

We model the stochastic nexus between the arrival of a possibly injurious alien species and the attendant monetary damage that results if this species is able to establish itself in the new habitat of Home. In this regard, we shall say that the monetary damages associated with the possibly successful introduction of alien species "arrive" at Home in accordance with a Poisson process with rate λ and λ= λ(*m*). In other words, the arrival rate of the monetary damages stemming from the stochastic introduction of alien species is a function of the volume of imports *m*. Further, we suppose that as the volume of imports goes up, the arrival rate of the monetary damages also goes up. Therefore, we have $\lambda^{/}(m)>0$. The *amounts* of the successive monetary damages are independent random variables that are assumed to have the common discrete distribution $a_k = -\pi^k[klog_e(1-\pi)]^{-1}$, where *k*=1,2,3,... and the parameter $\pi \in (0,1)$. Our task now is to ascertain the distribution of the total monetary damage in an interval $[0,t]$ and, without loss of generality, we suppose that this interval is a calendar year.

Let us now compute the generating function of the discrete probability distribution specified in the previous paragraph.[20] Because the common discrete distribution $a_k = -\pi^k[k\log_e(1-\pi)]^{-1}$, k=1,2,3,... has the natural logarithm function in it, to make further progress, we want to work with a series expansion of the natural logarithm function. This series expansion is given by

$$-\log_e(1-x)=\sum_{k=0}^{\infty}\frac{x^k}{k},\quad |x|<1. \tag{1}$$

Further, Theorem 1.2.1 in Tijms (2003, p. 19) tells us that the generating function of the total monetary damage in a year can be written in terms of the exponential *e*. Now, using this result from Theorem 1.2.1, the series expansion in equation (1), and some thought, we are able to conclude that the generating function of the total monetary damage in a year is given by

20See Kulkarni (1995, pp. 579-584) and Tijms (2003, pp. 449-455) for textbook accounts of the generating function.

$$e^{-\lambda\{1-A(z)\}},\ A(z)=\frac{\log_e(1-\pi z)}{\log_e(1-\pi)},\ |z|\leq 1. \tag{2}$$

After several algebraic steps, the generating function in equation (2) can be written as

$$[\frac{(1-\pi)}{(1-\pi z)}]^{-\lambda/\log_e(1-\pi)},\ |z|\leq 1. \tag{3}$$

Consulting Kulkarni (1995, p. 584), it is clear that the generating function in equation (3) is the generating function of a random variable that has a negative binomial distribution with parameters $-\lambda/\log_e(1-\pi)$ and $(1-\pi)$. Therefore, we reason that the total monetary damage from biological invasions in a calendar year has a negative binomial distribution with the above specified parameters.

Using standard formulas for the negative binomial distribution, we can tell that the expected total monetary damage from biological invasions in a calendar year or $E[D_1]$ is

$$E[D_1]=\frac{-\pi\lambda}{\{(1-\pi)\log_e(1-\pi)\}}. \tag{4}$$

Equation (4) gives us our first measure of the damage from stochastic alien species introductions in Home. The right-hand-side (RHS) of equation (4) is positive because the numerator and the denominator on the RHS are both negative. The reader will note that the expected total monetary damage from biological invasions in a calendar year is given by a particular ratio. The numerator of this ratio is the (negative) product of the rate λ of the Poisson arrival process of the monetary damages and the parameter π of the discrete distribution function of the amounts of the consecutive monetary damages. The denominator is the product of a simple function of the parameter π, for example, $(1-\pi)$, and the natural logarithm of this same function. Inspection of the above expression for the expected value tells us that as the rate λ of the Poisson arrival process increases in magnitude, the expected total monetary damage from biological invasions goes up. We now compute our second measure of the monetary damage from alien species introductions.

2.3. Second measure of damage

As in section 2.2, we wish to compute the *expected* total monetary damage in the interval $[0,t]$ which, without loss of generality, is a calendar year. Once again, we assume that alien species are introduced into Home in accordance with a Poisson process with rate λ, where $\lambda=\lambda(m)$ and $\lambda'(m)>0$. In words, the rate of introduction of alien species into Home, λ, is a function of the volume of imports m and this introduction rate is an increasing function of the volume of imports.[21] The *ith* introduction causes monetary damage M_i, $i \geq 1$. The M_i are assumed to be independent and identically distributed (i.i.d.), and they are also assumed to be independent of the total number of alien species introductions by time *t*, *N*(*t*). As one might expect, the monetary damage from a specific introduction in the calendar year [0,*t*] is typically not constant, but variable. Therefore, we suppose that the monetary damage from a specific alien species introduction decreases exponentially over time. Mathematically, this means that if the initial damage from a particular species introduction is *M*, then at some later time *t* the damage is Me^{-at}, where α is the parameter or the rate of the exponential distribution.

With the above description in place, we can now tell that the total monetary damage from invasive species introductions into Home in a calendar year is

$$D_2 = \sum_{i=1}^{N(t)} M_i e^{-\alpha(t-A_i)}, \tag{5}$$

21 These assumptions have also been made by Costello and McAusland (2003). For more on the Poisson process, the reader should consult Ross (1996, chapter 2) or Tijms (2003, chapter 1).

where A_i is the arrival time of the *ith* introduction. Obviously, D_2 is a random variable. Therefore, let us now compute $E[D_2]$, the expected dollar damage from alien species introductions into Home in a calendar year. Conditioning on $N(t)$, the total number of introductions by time t, we get

$$E[D_2]=\sum_{j=0}^{\infty} E\left[D_2 / N(t)=j\right] e^{-\alpha t}\{\frac{\alpha t)^j}{j!}\}. \tag{6}$$

Theorem 2.3.1 in Ross (1996, p. 67) tells us that conditioned on $N(t)=j$, the unordered arrival times $A_1,\ldots A_j$, are distributed as independent, uniform random variables in the interval $[0,t]$. From this, we deduce that given $N(t)=j$, D_2 has the same distribution as $\sum_{i=1}^{j} M_i e^{-\alpha(t-G_i)}$, where the $G_i, i=1,\ldots j$ are independent and uniformly distributed random variables in $[0,t]$. Putting these last two pieces of information together, we get

$$E\left[D_2 / N(t)=j\right]=jE[M]E[e^{-\alpha(t-G)}], \tag{7}$$

where $E[M]$ is the initial monetary damage caused by a particular introduction and G is a uniformly distributed random variable in $[0,t]$.

To compute the last expectation on the RHS of equation (7), observe that

$$E[e^{-\alpha(t-G)}]=\int_0^t \frac{1}{t}e^{-\alpha(t-g)}dg=(\frac{1-e^{-\alpha t}}{\alpha t}). \tag{8}$$

Using equation (8), we can now write

$$E[D_2 / N(t)]=N(t)E[M](\frac{1-e^{-\alpha t}}{\alpha t}). \tag{9}$$

Finally, taking expectations and using the fact that $E[N(t)]=\lambda t$, we obtain

$$E[D_2]=(\frac{\lambda E[M]}{\alpha})(1-e^{-\alpha t}). \tag{10}$$

Equation (10) gives us our second measure for the expected monetary damage in Home from the stochastic introduction of invasive species in the interval $[0,t]$ that is a calendar year.

This expected monetary damage depends on the mean initial monetary damage from an introduction ($E[M]$), on the rate (α) of the exponential distribution, and most importantly for our purpose, on the rate (λ) at which alien species are being introduced into Home. Recall that because $\lambda=\lambda(m)$, the expected monetary damage metric given by equation (10) also depends on the volume of imports (m) coming into

Home. We now proceed to compute our third and final measure of the damage from alien species introductions. This metric is a physical measure of damage.

2.4. Third measure of damage

Upon arrival in Home, ships unload their containers carrying cargo. The arrival of these containers (and the possible discharge of ballast water) coincides with the arrival of a whole host of potentially deleterious alien animal and plant species. Consistent with the approach adopted in sections 2.2 and 2.3, we suppose that the arrival process of these alien species can be described with a Poisson process with rate λ[22]. What's different in our construction of this third measure of damage is that, unlike most of the previous literature on this subject, we suppose that this rate λ of the Poisson arrival process is a *random variable* that follows a gamma distribution with shape parameter α and scale parameter β. [23] As in sections 2.2 and 2.3, we assume that the time

22See Costello and McAusland (2003), Batabyal and Nijkamp (2005), and Batabyal and Lee (2006) for additional details on this point.

23See Ross (1996, p. 18) and Tijms (2003, pp. 441-442) for more on the gamma distribution. We are using the gamma distribution to characterize the rate λ for four reasons. First, this distribution has been used previously in the natural resource economics literature (see Batabyal and Nijkamp 2009) to study stochastic arrival processes. Second, the gamma distribution is a general two parameter distribution for positive random variables. Third, many other probability distribution functions are variants of the gamma distribution. Finally, in Bayesian inference, the conjugate prior of the unknown rate parameter λ is commonly modeled with the gamma distribution.

interval of interest $[0,t]$ is a calendar year. The task before us now is to compute the total number of biological invasions in this calendar year.

Let p_k denote the probability that there are k biological invasions in a year. Then, using the law of conditional probabilities and the fact that the rate λ of the alien species arrival process is gamma distributed with parameters (α,β), we reason that

$$p_k = \int_0^\infty e^{-x} \left(\frac{x^k}{k!}\right) \beta^\alpha \left(\frac{x^{\alpha-1}}{\Gamma(\alpha)}\right) e^{-\beta x}\, dx, \tag{11}$$

where $\Gamma(\alpha)$ is the gamma function.

Now, we know that the gamma density function $\gamma^\delta\{x^{\delta-1}/\Gamma(\delta)\}$ $e^{-\gamma-x}$ integrates to unity over the interval $(0,\infty)$ for any γ, $\delta > 0$. Therefore, after several steps of algebra, we see that

$$p_k = \frac{\Gamma(\alpha+k)}{\Gamma(\alpha)\Gamma(k+1)} \left(\frac{\beta}{1+\beta}\right)^\alpha \left(\frac{1}{1+\beta}\right)^k, \; k=0,1,2,\ldots \tag{12}$$

Consulting Ross (1996, p. 16), it is clear that equation (12) describes the probability mass function of a negative binomial random variable with parameters α and $\beta/(\beta+1)$. Therefore, the total number of biological invasions in a calendar year has a negative binomial distribution with parameters α and $\beta/(\beta+1)$.

Using standard formulas for the negative binomial distribution, it is straightforward to confirm that the expected number of biological invasions in a calendar year or $E[D_3]$ is

$$E[D_3]=\alpha/\beta. \tag{13}$$

Equation (13) gives us our third and final measure of damage from stochastic alien species introductions in Home. Specifically, this equation tells us that the expected number of biological invasions in a calendar year is given by the ratio of the shape parameter α of the gamma distribution to the scale parameter β of this same distribution. As α increases in magnitude, the expected number of biological invasions goes up, and as β increases in magnitude, the expected number of biological invasions goes down. We now move on to analyze the effects of small and optimal tariffs imposed by Home on the expected damage from invasive species, on prices, on exports and imports, and on social welfare in Home for four alternate market structures.

3. Tariffs and Alien Species Management

3.1. Tariffs and social welfare in Home

Our two country model is adapted from the standard two country trade model discussed in Feenstra (2004, chapter 7). Home imports a single (agricultural or manufactured) good from Foreign. The price of the import good in Home is p and its world price is p^*.[24] To keep the subsequent analysis straightforward and to avoid focusing on too many cases, we shall analyze the effects of a Home instituted *specific* import tariff τ.[25] Given τ, it is clear that $p = p^* + \tau$. In addition to the import good, we suppose that the second (numeraire) good is also traded at the fixed world price of unity. Labor (L) is the only factor of production, and we suppose that each unit of the numeraire good requires one unit of labor. Therefore, wages in Home are also unity, and hence total labor income in Home equals the fixed supply of labor L. The reader will note that we are in a partial equilibrium setting in which wages are fixed and trade is balanced through flows of the numeraire good.

24In the rest of this chapter we shall adopt the convention of denoting the relevant Foreign functions and variables with the * superscript.

25Results for an *ad valorem* tariff can be expected to be qualitatively similar to that for a specific tariff.

The output of the good in question in Home is q, and the industry cost of producing this good is $c(q)$ where $c^{/}(q) > 0$ and $c^{//}(q) > 0$. Imports are denoted by the scalar $m=d(p)-q$, where $d(p)$ is the demand function for the good whose output is q, and we suppose that $d^{/}(p) < 0$. The tariff revenue τm is returned to the citizens of Home, and these citizens also obtain the profits of the import competing industry $pq-c(q)$. Given this specification, we can now write the social welfare function in Home at time t as

$$W\{p, L + \tau m + pq - c(q), E[D_i]\} \equiv W(\tau), \; i=1,2,3. \tag{14}$$

We see that social welfare in Home depends on prices (p), income ($L+\tau m+pq-c(q)$), and the expected total damage (see equations (4), (10), and (13)) from the probabilistic introduction of alien species ($E[D_i]$), i=1,2,3. The social welfare function described by equation (14) is like an indirect utility function for Home as a whole, with one caveat:[26] Home's welfare at time t also depends *negatively* on the expected damage from the introduction of alien species.

We are now in a position to analyze the impact of the specific tariff on prices, quantities, mean damage, and social welfare under

26For additional details on this point see Feenstra (2004, pp. 213-214).

alternate market structures. To this end, let us first derive a general expression for the change in social welfare when a tariff is put in place by Home. The derivation of this general expression will be helped by first noting that

$$\frac{dE[D_1]}{d\tau}=\frac{-\pi}{(1-\pi)\log_e(1-\pi)}\frac{d\lambda}{dm}\frac{dm}{dp}\frac{dp}{d\tau},\ \frac{dE[D_2]}{d\tau}=$$

$$\frac{E[M](1-e^{-\alpha t})}{\alpha}\frac{d\lambda}{dm}\frac{dm}{dp}\frac{dp}{d\tau},\ \frac{dE[D_3]}{d\tau}=0 \tag{15}$$

Continuing with the derivation, we assume that p and q both depend on the tariff τ. Totally differentiating equation (14) with respect to τ, we get

$$\frac{dW}{d\tau}= -d(p)\frac{dp}{d\tau}+m+\{\tau\frac{dm}{dp}+q+Z_i\frac{\partial W}{\partial(E[D_i])}\frac{d\lambda}{dm}\frac{dm}{dp}\}\frac{dp}{d\tau}+$$

$$\{p-c'(q)\}\frac{dq}{d\tau},\ i=1,2,3, \tag{16}$$

where

$$Z_1=[-\pi/\{(1-\pi)\log_e(1-\pi)\}]>0,\ Z_2=[\{E[M](1\ e^{-\alpha t})\}/\alpha]>0,\ Z_3=0, \tag{17}$$

and we have used the fact that $\partial W/\partial p = -d(p)$. Now, note that $m=d(p)-q$, and because $p=p^*+\tau$, we have $\{1-dp/d\tau\}=-dp^*/d\tau$. Using these two pieces of information, we can simplify equation (16). This simplification yields

$$\frac{dW}{d\tau}=\{\tau\frac{dm}{dp}+Z_i\frac{\partial W}{\partial(E[D_i])}\frac{d\lambda}{dm}\frac{dm}{dp}\}\frac{dp}{d\tau}$$

$$-m\frac{dp^*}{d\tau}+\{p-c'(q)\}\frac{dq}{d\tau},\ i=1,2,3 \qquad (18)$$

Let us examine the three terms on the RHS of equation (18) in greater detail. The first term can be thought of as the efficiency effect of the tariff. The second term is the effect of the tariff on the foreign price p^* or the terms of trade effect. Finally, the third term is the price-cost margin multiplied by the change in the industry output. In the remainder of this chapter, we shall use equation (18) repeatedly to study the effects of the specific tariff (τ) under alternate market structures. We begin with the case in which perfect competition prevails in both Home and Foreign and Home is a small country.

3.2. Perfect competition with Home a small country

When the home economy is perfectly competitive and it is a small country, we have $p=c'(q)$ and because p^* is fixed, we also have

$dp^*/d\tau = 0$ and $dp/d\tau = 1$. We now simplify equation (18) using these three results. This gives us

$$\frac{dW}{d\tau} = \tau\frac{dm}{dp} + Z_i\frac{\partial W}{\partial(E[D_i])}\frac{d\lambda}{dm}\frac{dm}{dp},\ i=1,2,3, \tag{19}$$

and the Z_i in equation (19) are given in equation (17). Now recall from the discussion in section 3.1 that the expected damage from alien species introductions influences social welfare in Home negatively, and hence $\partial W/\partial(E[D_i]) < 0$. We have already noted in section 2 that increasing the volume of imports increases the rate of introductions, and therefore $d\lambda/dm > 0$. Finally, $dm/dp = d'(p) - \{1/c''(q)\}$. Because the demand function slopes downward and the cost function is strictly convex, we have $d'(p) < 0$ and $c''(q) > 0$ and hence $dm/dp < 0$. We now use these three results and evaluate equation (19) at $\tau = 0$. This gives us

$$\frac{dW}{d\tau}\Big|_{\tau=0} = Z_i\frac{\partial W}{\partial(E[D_i])}\frac{d\lambda}{dm}\frac{dm}{dp} > 0,\ i=1,2,\ \frac{dW}{d\tau}\Big|_{\tau=0} = 0,\ i=3. \tag{20}$$

Equation (20) tells us that when the expected damage in Home from stochastic alien species introductions is *monetary* (i.e., when equations (4) and (10) are relevant), starting from a position of free trade ($\tau = 0$), a *small* tariff unambiguously raises social welfare in Home. In contrast, when the expected damage from stochastic alien species introductions is *physical* (i.e., when equation (13) is pertinent), starting from a position of free trade, a small tariff does not raise welfare. This last result arises because the expected total physical damage (see equations (13), (15), and (17)) is independent of the rate λ of the Poisson arrival process. Put differently, the small tariff has no impact on the expected total damage from alien species introductions, and hence this tariff also has no impact on social welfare in Home. Given these results, we now determine the impact of an *optimal* tariff on social welfare in Home. To compute the optimal tariff, we set the RHS of equation (19) equal to zero and then simplify. This gives us

$$\tau_o = -Z_i \frac{\partial W}{\partial(E[D_i])}\frac{d\lambda}{dm} > 0,\ i=1,2,\ \tau_o = 0,\ i=3. \tag{21}$$

Equations (20) and (21) together tell us that when social welfare depends on the expected damage from the introduction of alien species, small and optimal tariffs both *raise* welfare in Home as long as the expected total damage in Home is monetary (i.e., when equations (4) and (10) are relevant). In contrast, when the expected total damage is physical and independent of the rate λ, a tariff is incapable of affecting social welfare in Home, and hence, in this last case, it is optimal to not use a tariff to regulate alien species in Home. Even though Home is a small country, we have seen that in two out of the three cases that we are studying, tariffs have a positive impact on social welfare, and hence Home ought to have an activist trade policy in place in these two cases.

Stepping away from the criterion of social welfare for a moment, does a small tariff lower the expected total damage from the introduction of alien species? To answer this question, let us inspect equation (15) carefully. This inspection tells us that $dE[D_1]/d\tau < 0$, $dE[D_2]/d\tau < 0$, $dE[D_3]/d\tau = 0$. In this chapter, imports are the only means by which alien species are introduced into Home. As a result, because a small tariff reduces the volume of imports when the expected total damage is monetary (i.e., $dm/d\tau < 0$), this same tariff also *lowers* the expected monetary damage in Home from the introduction of alien species.[27] In contrast, when the expected total damage is physical and independent of the rate λ, a small tariff has no impact on the expected total damage. This is what the derivative $dE[D_3]/d\tau = 0$ in equation (15) is telling us.

It is well known in international trade theory (see Feenstra 2004, p. 216) that the optimal tariff for a small country is zero. However, our analysis thus far tells us that this result does not hold in *some cases* in which imports and invasive species go together. In fact, as we have just seen, when imports are the only means by which alien species are

27Using a different model, Costello and McAusland (2003) have obtained a similar result.

introduced into Home, and the expected total damage from alien species is given by either equation (4) or equation (10), the optimal course of action for Home is to set a positive tariff. We now investigate the effects of a tariff when Home is a large country.

3.3. Perfect competition with Home a large country

We now write the world price of imports as $p^*(\tau)$. Therefore, because $p = p^*(\tau) + \tau$, $dp^*/d\tau \neq 0$ and $dp/d\tau \neq 1$. However, because the Home economy is perfectly competitive, we still have $p = c'(q)$. Using these three results to simplify equation (18), we get

$$\frac{dW}{d\tau} = \{\tau\frac{dm}{dp} + Z_i\frac{\partial W}{\partial(E[D_i])}\frac{d\lambda}{dm}\frac{dm}{dp}\}\frac{dp}{d\tau} - m\frac{dp^*}{d\tau}, \quad i=1,2,3. \qquad (22)$$

Now, because Home is a *large* country, $dp/d\tau \neq 1$, and we expect that the foreign exporter will absorb a part of the Home tariff. As noted in Feenstra (2004, pp. 218-219), this means that in general, we expect $dp^*/d\tau < 0$ and $0 < dp/d\tau < 1$. Let us now use these two findings and the results stated in section 3.2 to evaluate equation (22) at $\tau = 0$. This gives us

$$\frac{dW}{d\tau}\Big|_{\tau=0}=Z_i\frac{\partial W}{\partial(E[D_i])}\frac{d\lambda}{dm}\frac{dm}{dp}\frac{dp}{d\tau}-m\frac{dp^*}{d\tau}>0,\ i=1{,}2,\ \frac{dW}{d\tau}\Big|_{\tau=0}=$$

$$-m\frac{dp^*}{d\tau}>0,\ i=3. \tag{23}$$

Equation (23) tells us two things of note. First, as in section 3.2, we see that, when the expected total damage from alien species is monetary (equations (4) and (10) apply), starting from a position of free trade, a small tariff, once again, *raises* social welfare in Home. Second, and unlike what we saw in section 3.2, when the expected total damage from alien species is physical and independent of λ, the small tariff now is not zero, but positive. This is because, even though the small tariff is unable to affect the expected total damage from alien species, because Home is a large country, the small tariff is able to generate a beneficial terms of trade effect and hence, as shown by the $-m(dp^*/d\tau)$ term in equation (23), this tariff is positive.

To compute the optimal tariff for Home, we set the RHS of equation (22) equal to zero and then simplify the resulting expression. This gives us

$$\tau_o=\frac{m\frac{dp^*}{d\tau}}{\frac{dm}{dp}\frac{dp}{d\tau}}-Z_i\frac{\partial W}{\partial(E[D_i])}\frac{d\lambda}{dm}>0,\ i=1{,}2,\ \tau_o=\frac{m\frac{dp^*}{d\tau}}{\frac{dm}{dp}\frac{dp}{d\tau}}>0,\ i=3. \tag{24}$$

Equation (24) tells us that for the case of monetary expected total damage from alien species (i.e., for i=1,2), Home's optimal tariff is positive and is the sum of the terms of trade effect and the damage from alien species effect. In contrast, for the physical damage or the i=3 case, there is no damage from alien species effect to contend with because Z_3 =0 but, unlike the case studied in section 3.2, there is still a beneficial terms of trade effect. This is why the optimal tariff in this i=3 case is also positive. The reader will note that the optimal tariff in this i=3 case is equivalent to the optimal import tariff for a large country in the *absence* of damage from alien species.[28] Equation (24) also tells us that, when the damage from invasive species introductions *is* an issue (i.e., when i=1,2), the optimal tariff is not only positive, but also larger in magnitude than the optimal tariff with no invasive species damage.

The impact of a small tariff on the expected total damage from alien species introductions into Home is, once again, given by equation (15). Although $dp/d\tau \neq 1$ now, as discussed earlier, in general, we expect $dp/d\tau > 0$ to hold. Therefore, inspection of equation (15) and some thought together tell us that $dE[D_1]/d\tau < 0$, $dE[D_2]/d\tau < 0$, $dE[D_3]/d\tau = 0$. Since a small tariff reduces the volume of imports when the expected total damage is monetary (i.e., $dm/d\tau < 0$) this same tariff also

28See Feenstra (2004, p. 220) for more details on this point.

reduces the expected monetary damage in Home from the introduction of alien species. In contrast, when the expected total damage is physical and independent of λ, a small tariff has no impact on the expected total damage, and therefore $dE[D_3]/d\tau = 0$. Let us now analyze the effects of tariffs when the exporter in Foreign is a monopolist and there are no import competing firms in Home.

3.4. Monopolist in Foreign

In this case, we have a single exporter in Foreign, and we suppose that there are no import competing firms in Home. The purpose of a tariff is generally to protect domestic firms in Home, so if there are no domestic firms, then, in principle, there is no rationale for an import tariff. However, as we shall see, in our case, it is the damage from alien species introductions that provides a rationale for protectionism.

Let us denote the Foreign firm's exports to Home by x; this equals Home consumption and therefore we can write $x=d(p)$. Inverting this expression, we get the inverse demand function $p=p(x)$ where $p^{/}(x)<0$. Denote the price received by the Foreign exporter by $p^{*}=p(x)-\tau$, and let $c^{*}(x)$ denote this firm's cost function. We suppose that $c^{*'}(x)>0$ and that $c^{*//}(x)>0$. The Foreign exporter's profit function is $\pi^{*}(x)=x\,p^{*}-c^{*}(x)=x\{p(x)-\tau\}-c^{*}(x)$. Maximizing this function with respect to the volume of exports x gives us

$$\pi^{*/}(x)=p(x)+x\,p^{/}(x)-\{c^{*/}(x)+\tau\}=0, \tag{25}$$

where $p(x)+x\,p^{/}(x)$ is the marginal revenue and $\{c^{*/}(x)+\tau\}$ is the tariff inclusive marginal cost. Now, totally differentiating equation (25), we get

$$\frac{dx}{d\tau}=\frac{1}{2p^{/}(x)+xp^{//}(x)-c^{*//}(x)}=\frac{1}{\pi^{*//}(x)}<0 \tag{26}$$

and

$$\frac{dp}{d\tau}=\frac{dp}{dx}\frac{dx}{d\tau}=\frac{p^{/}(x)}{\pi^{*//}(x)}>0. \tag{27}$$

Equation (26) tells us that the Foreign firm's exports decline as a result of the tariff, and equation (27) tells us that the domestic price of the good in question inclusive of the tariff rises.

When $dp/d\tau=1$ (as in section 3.2), the so called "pass-through" of the tariff is complete. In other words, the tariff inclusive price p rises by the same amount as the tariff τ. However, when Home is not a small country, the "pass-through" of the tariff will typically be incomplete. When this is the case, we will have $dp/d\tau<1$. This also means that the foreign firm will absorb a part of the tariff. Mathematically, this means that $d p^{*}/d\tau < 0$. It should be clear to the reader that, when the pass-through of a tariff is *incomplete*, there is a terms of trade *gain* for Home.

We would now like to derive a condition that tells us when there will be a terms of trade gain for Home. Equation (27) helps to provide us with the answer. Because the numerator and the denominator on the RHS of equation (27) are both negative, we conclude that $dp/d\tau < 1$ if and only if

$$p'(x) > 2p'(x) + xp''(x) - c^{*''}(x). \tag{28}$$

Now, following the discussion in Feenstra (2004, pp. 221-222), we can say that the LHS of the inequality in (28) is the slope of the inverse demand function, and the RHS of this inequality is the difference between the slopes of the marginal revenue and the marginal cost functions. To proceed further with the derivation, it will be helpful to suppose that the Foreign cost function $c^{*}(x)$ is linear. Then $c^{*''}(x)=0$ and the inequality in (28) reduces to

$$p'(x) > 2p'(x) + xp''(x). \tag{29}$$

The inequality in (29) and some thought together tell us that when the slope of the marginal revenue function *exceeds* that of the inverse demand function, $dp/d\tau < 1$ and $d p^{*}/d\tau < 0$.

We now determine the impact of the tariff on social welfare in Home. Because there are no import competing firms in Home, q=0, and therefore $dq/d\tau$ =0. Using this result to simplify equation (18), we get

$$\frac{dW}{d\tau}=\{\tau\frac{dm}{dp}+Z_i\frac{\partial W}{\partial(E[D_i])}\frac{d\lambda}{dm}\frac{dm}{dp}\}\frac{dp}{d\tau}-m\frac{dp^*}{d\tau},\ i=1,2,3. \tag{30}$$

From the discussion in the previous paragraph, we know that when the marginal revenue function is steeper than the inverse demand function, $dp/d\tau<1$ and $dp^*/d\tau<0$. Using these two results and other results from our earlier analysis, let us evaluate equation (30) at τ =0. We get

$$\frac{dW}{d\tau}\Big|_{\tau=0}=Z_i\frac{\partial W}{\partial(E[D_i])}\frac{d\lambda}{dm}\frac{dm}{dp}\frac{dp}{d\tau}-m\frac{dp^*}{d\tau}>0,\ i=1,2,\quad \frac{dW}{d\tau}\Big|_{\tau=0}=-m\frac{dp^*}{d\tau}>0,\ i=3. \tag{31}$$

We see that when the expected total damage from alien species is monetary (i.e., when equations (4) and (10) apply and i=1,2), starting from a position of free trade, a small tariff *raises* social welfare in Home. In addition, when the expected damage from alien species is physical and independent of the rate λ (i.e., when i=3 as in section 3.3), a beneficial terms of trade effect results from the small tariff and this explains why this tariff is, as shown by the last derivative in equation (31), positive.

Does the optimal tariff also raise welfare? To answer this question, we set the RHS of equation (30) equal to zero and then simplify the resulting expression. This gives us

$$\tau_o = \frac{x\frac{dp^*}{d\tau}}{\frac{dx}{d\tau}} - Z_i \frac{\partial W}{\partial\left(E\left[D_i\right]\right)} \frac{d\lambda}{dm} > 0,\ i=1,2,\ \ \tau_o = \frac{x\frac{dp^*}{d\tau}}{\frac{dx}{d\tau}} > 0,\ i=3. \quad (32)$$

The optimal import tariff when the exporter in Foreign is a monopolist, when there are no import competing firms in Home, and when the expected total damage from alien species is physical and independent of the rate λ is given by the second expression in (32). This expression is the ratio of two negative quantities, and hence the optimal tariff described by this expression is positive. Equation (32) tells us that when the damage from invasive species introductions is monetary, we have $Z_i > 0$, i=1,2, and in these two instances, the optimal tariff, given by the first expression in (32), is not only positive, but also larger in magnitude than the optimal tariff with physical alien species damage.

As in sections 3.2 and 3.3, equation (15) gives us the impact of a small tariff on the expected total damage from alien species introductions into Home. Equation (27) tells us that $dp/d\tau > 0$. Using this and our previous results in equations (15) and (17) tells us that $dE[D_1]/d\tau < 0$, $dE[D_2]/d\tau < 0$, $dE[D_3]/d\tau = 0$. Since a small tariff reduces the volume of imports when the expected total damage is monetary (i.e., $dm/d\tau < 0$), this same tariff also *reduces* the expected monetary damage in Home from the introduction of alien species. In contrast, when the expected total damage is physical and independent of λ, a small tariff has no impact on the expected total damage, and hence $dE[D_3]/d\tau = 0$. We now study the impact of a tariff when the Foreign exporter and an import competing firm in Home engage in Cournot competition.

3.5. Cournot competition

We now have a Home (domestic) firm competing with an exporting firm from Foreign in the domestic market. Let us denote the sales of the Foreign exporting firm by *x* and that of the Home import competing firm by *q* so that aggregate consumption of the good in question at Home is $y=q+x$. Following the logic of section 3.4, the pertinent demand function now is $y=d(p)$, and therefore the relevant inverse demand function is $p=p(y)$ where $p'(y) < 0$.

Using the notation of section 3.4, the profit functions of the Foreign exporter and the Home import competing firm are $\pi^{*}(x)=x\{p(y)-\tau\}-c^{*}(x)$ and $\pi(q)=qp(y)-c(q)$. Maximizing these two functions with respect to the choice variables x and q respectively, we get

$$\pi^{*/}(x)=p(y)+x\,p^{/}(y)-\{c^{*/}(x)+\tau\}=0 \tag{33}$$

and

$$\pi^{/}(q)=p(y)+q\,p^{/}(y)-c^{/}(q)=0 \tag{34}$$

The reader can confirm that the two second order conditions for profit maximization are $\pi^{*//}(x)=2\,p^{/}(y)+x\,p^{//}(y)-c^{*//}(x)<0$ and $\pi^{//}(q)=2\,p^{/}(y)+q\,p^{//}(y)-c^{//}(q)<0$.[29] We now want to determine the impact of the Home tariff on the Foreign firm's exports. In other words, we want to determine the sign of the derivative $dx/d\tau$. To determine this sign, we totally differentiate equation (33) and then use the second order condition for $\pi^{*//}(x)$ given above. This gives us

29 We assume that the stability condition $\pi^{*}_{xx}\pi_{qq}-\pi^{*}_{xq}\pi_{qx}>0$ is satisfied.

$$\frac{dx}{d\tau}=\frac{1}{2p^{/}(y)+xp^{//}(y)-c^{*//}(x)}=\frac{1}{\pi^{*//}(x)}<0. \tag{35}$$

Equation (35) tells us that with the Home tariff in place, the Foreign exporter reduces the amount it wishes to export to Home.

To study the impact of the Home tariff on prices, it will be necessary first to compute the impact of the tariff on total sales $y=x+q$. To do this, let us now sum the two first order necessary conditions given in equations (33) and (34). This gives

$$2p(y)+y\,p^{/}(y)=c^{/}(q)+\{c^{*/}(x)+\tau\}. \tag{36}$$

Totally differentiating equation (36) and then simplifying, we get

$$3\,p^{/}(y)dy+y\,p^{//}(y)dy=c^{//}(q)dq+c^{*//}(x)dx+d\tau. \tag{37}$$

Now, as in section 3.4, to progress further it will help to make a simplifying assumption. Therefore, we assume that the cost functions of the two competing firms in Home and in Foreign are both linear. Then, equation (37) can be simplified to give us

$$\frac{dy}{d\tau}=\frac{1}{3p^{/}(y)+yp^{//}(y)}, \tag{38}$$

and, because $dp/d\tau$ =(dp/dy)(dy/$d\tau$), we have

$$\frac{dp}{d\tau}=\frac{p^{/}(y)}{3p^{/}(y)+yp^{//}(y)}. \tag{39}$$

Inspecting the denominators on the RHSs of equations (38) and (39), we see that the impact of the Home tariff on total output (y) and the price (p) depends significantly on the sign of $3p^{/}(y)+yp^{//}(y)$. When this expression is negative, we have $dy/d\tau<0$ and $dp/d\tau>0$. In words, total output with the tariff declines, and the price in Home with the tariff rises. Some thought will convince the reader that the condition $3p^{/}(y)+yp^{//}(y)<0$ holds for some inverse demand functions (such as the linear function), but not for all such functions. This tells us that the imposition of a tariff by Home *may* lead to counterintuitive results. Specifically, total output with the tariff may rise ($dy/d\tau>0$), and the domestic price of the good in question at Home may fall ($dp/d\tau<0$).

We now focus on the "pass-through" of the tariff τ. We know that for there to be a terms of trade gain in Home, we must have $0<$ dp/d

$\tau < 1$. Now, when $3p'(y)+yp''(y)<0$ holds, from equation (39) we can tell that for $dp/d\tau < 1$ to hold, we must have

$$p'(y) > 3p'(y)+yp''(y) \Leftrightarrow 0 > 2p'(y)+yp''(y). \quad (40)$$

The expression $2p'(y)+yp''(y)$ on the RHS of (40) is the slope of the marginal revenue function $p(y)+y\,p'(y)$. Therefore, what (40) is really saying is that when the marginal revenue function is *downward* sloping $dp/d\tau < 1$ holds or, alternately, $d\,p^*/d\tau < 0$, and hence there is a terms of trade gain for Home.

We now ascertain the impact of the tariff on social welfare in Home. Because $m=x$ in equilibrium, we use this to rewrite equation (18) as

$$\frac{dW}{d\tau} = \{\tau\frac{dx}{dp} + Z_i \frac{\partial W}{\partial(E[D_i])}\frac{d\lambda}{dx}\frac{dx}{dp}\}$$

$$\frac{dp}{d\tau} - x\frac{dp^*}{d\tau} + \{p - c'(q)\}\frac{dq}{d\tau}, \; i=1,2,3. \quad (41)$$

From equation (35) we know that $dx/d\tau < 0$. Further, as we have just discussed in the previous paragraph, when the inequality in (40) holds, the marginal revenue function is downward sloping, and hence *d*

$p^*/d\tau<0$. Finally, because $dq/d\tau=(dq/dp)(dp/d\tau)$, in general, we expect $dq/d\tau>0$. [30] Using these three results we can evaluate equation (41) at $\tau=0$. This gives us

$$\frac{dW}{d\tau}|_{\tau=0}=Z_i\frac{\partial W}{\partial(E[D_i])}\frac{d\lambda}{dx}\frac{dx}{d\tau}-x\frac{dp^*}{d\tau}+\{p\text{-}c'(q)\}\frac{dq}{d\tau}>0,\ i=1,2,$$

$$\frac{dW}{d\tau}|_{\tau=0}=-x\frac{dp^*}{d\tau}+\{p\text{-}c'(q)\}\frac{dq}{d\tau}>0,\ i=3. \tag{42}$$

Consider the cases in which the expected total damage from alien species introductions is monetary. These are the i=1,2 cases. In these two cases, equation (42) tells us that when the inequality in (40) holds, a small tariff leads to a terms of trade gain ($dp^*/d\tau<0$). In addition, when $dq/d\tau>0$, there is an additional gain from this small tariff. These two positive effects, along with the fact that the first term on the RHS of equation (42) with Z_i in it is positive together tell us that a small tariff raises social welfare in Home. Next, consider the i=3 case in which the expected total damage from alien species introductions is physical and independent of λ. In this case Z_3=0, but we still have a beneficial terms of trade effect and a general increase in domestic output, as a result of

30See Feenstra (2004, pp. 225-226) for a graphical explanation of this line of reasoning.

the small tariff. These two positive effects explain why the small tariff in this i=3 case is also positive. The reader should note that in this case of Cournot competition, it is *not* inevitable that a small tariff will lead to an increase in domestic output. It is certainly possible that $dq/d\tau < 0$, and when this happens, this negative effect will tend to offset the positive terms of trade effect; hence the impact of a small tariff on social welfare in Home may not be positive.

Does the optimal import tariff raise social welfare in Home? To answer this question, we set the RHS of equation (41) equal to zero and then simplify the resulting expression. This gives us

$$\tau_o = \frac{x\frac{dp^*}{d\tau}}{\frac{dx}{d\tau}} - \frac{\{p(y)-c'(q)\}\frac{dq}{d\tau}}{\frac{dx}{d\tau}} - Z_i\frac{\partial W}{\partial(E[D_i])}\frac{d\lambda}{dx} > 0,\ i=1,2,\ \tau_o =$$

$$\frac{x\frac{dp^*}{d\tau}}{\frac{dx}{d\tau}} - \frac{\{p(y)-c'(q)\}\frac{dq}{d\tau}}{\frac{dx}{d\tau}} > 0,\ i=3. \tag{43}$$

Assuming that the inequality in (40) holds, let us first focus on the i=1,2 cases. There are three terms on the RHS of equation (43) to discuss. The first term is positive because both $dp^*/d\tau$ and $dx/d\tau$ are negative. The second term is positive because the numerator is generally positive and the denominator is negative. Finally, the third term is positive because $Z_i > 0$, i=1,2, $\partial W / \partial (E[D(t)]) < 0$, and $d\lambda/dx > 0$. Therefore, the optimal tariff, which is the sum of three positive terms, is itself positive. Next, focus on the i=3 case. In this case, the expected total damage from alien species introductions is independent of the rate λ, and hence this case is like the case in which there is Cournot competition between the exporting and the import competing firms and there is no damage from invasive species introductions. In this case, Home's optimal tariff is positive, and it is the sum of the first two terms as shown in the last derivative in equation (43). Finally, equation (43) tells us that the optimal tariff in the i=1,2 cases is the sum of three positive terms and hence larger in magnitude than the optimal tariff in the i=3 case.

As in sections 3.2, 3.3, and 3.4, equation (15) gives us the impact of a small tariff on the expected total damage from alien species introductions into Home. Using equation (35), the fact that $m=x$ in equilibrium, and our previous analysis, we reason that $dE[D_1]/d\tau < 0$, $dE[D_2]/d\tau < 0$, $dE[D_3]/d\tau = 0$. Since a small tariff diminishes the volume of exports when the expected total damage from alien species is monetary (i.e., $dx/ d\tau < 0$), this same tariff also *reduces* the expected monetary damage in Home from the introduction of alien species. In contrast, when the expected total damage is physical and independent of

λ, a small tariff has no impact on the expected total damage, and hence $dE[D_3]/d\tau = 0$. These results about the impact of the tariff on the expected total monetary damage from invasive species in sections 3.2 to 3.5 are similar to and consistent with Proposition 1 in Costello and McAusland (2003, p. 967).

In the strategic trade theory literature of the 1980s, a considerable amount of emphasis was placed on the third term $\{p-c'(q)\}(dq/d\tau)$ in equation (41), and this third term was often thought of as a profit shifting rationale for the strategic use of tariffs.[31] While this interpretation makes sense when the derivative $dq/d\tau$ is positive, we have already pointed out that this need not always be the case. Suppose for the moment that $dq/d\tau$ is positive. Then, what we have seen thus far in this chapter is that, in addition to any profit shifting rationale, in the presence of monetary damage from alien species introductions, there is a second and arguably more important rationale for the use of import tariffs. Indeed, when tariffs are used as described in this chapter, it may be possible to "kill three birds with one stone." What we mean by this expression is that the Home government may be able to (i) obtain a terms of trade benefit, (ii) shift profits away from the Foreign exporter and toward the domestic import competing firm, and (iii) reduce the

31 See Brander and Spencer (1984), Horstmann and Markusen (1986), and Helpman and Krugman (1989) for more on this literature.

monetary damage from deleterious alien species. We now discuss the form of the dependence of all the tariff expressions in this chapter on the expected total damage from alien species introductions in Home. Then, we briefly talk about scenarios in which this form of dependence would be different.

3.6. Form of dependence of tariffs on damage from alien species introductions

We derived three measures of damage from alien species introductions in sections 2.2 through 2.4 of this chapter. Of these three measures, the first two measures, equations (4) and (10), are monetary, and the third measure, equation (13), is physical. In our detailed analysis thus far, we have seen that tariffs are useful policies with which to control the deleterious effects of alien species introductions when equations (4) and (10) are pertinent (i.e., when the damage measure is monetary). This is because in these two cases, the rate λ of the Poisson arrival process directly influences the two derived damage metrics. In contrast, tariffs have no role to play as an alien species control device when equation (13) is germane because in this case, the derived damage metric is independent of λ.

We now note *two features* of our analysis thus far. First, even though equations (4) and (10) are *dissimilar* damage measures, in both these equations, the rate λ enters the damage measure multiplicatively.

Second, the three expected damage measures of this chapter enter the social welfare function in Home (see equation (14)) in a *standard* manner. In other words, we have $W\{p, L+\tau m+pq-c(\mathrm{q}), E[D_i]\}$, i=1,2,3. Therefore, when we differentiate this social welfare function with respect to the tariff τ, we get the multiplicative term $(\partial W/\partial\{E[D_i]\})(d\{E[D_i]\}/d\tau)$. It is these two modeling features that together account for the fact that the Z_i term affects all our tariff expressions (see equations (20), (21), (23), (24), (31), (32), (42) and (43)) multiplicatively.

We stress that the positivity of most of the tariffs that we have analyzed in this chapter is *not* the result of modeling the damage from alien species introductions in a particular way. In fact, as we have shown in this chapter, even for dissimilar damage measures, this positivity result largely holds. We conclude this section by pointing out that if the two modeling features delineated in the previous paragraph do not hold, then it is possible that the signs of some of the small and the optimal tariffs in the four different market structures that we have analyzed will become ambiguous.

4. Conclusions

In this chapter, we provided a theoretical perspective on the impacts of small and optimal specific tariffs when international trade in goods results in the stochastic introductions of alien species from one country to another as a byproduct. Conducting the analysis from the standpoint of the tariff imposing country (i.e., Home), we first derived three (two monetary and one physical) measures of the expected total damage from alien species introductions into Home. Next, we analyzed the effects of small and optimal tariffs under four alternate market structures. Our basic result is that there are several circumstances in which it makes sense to use trade policy (tariffs) to control the damage from alien plant and/or animal species.

The analysis in this chapter can be extended in a number of directions. In what follows, we suggest two possible extensions of this chapter's research. First, in our model, the rate of alien species introductions depends only on the volume of imports. Therefore, it would be useful to determine the extent to which one can obtain results from a model in which, in addition to the volume of imports, the rate of species introductions depends also on the *number of previously successful* introductions. Second, it would be useful to analyze the impacts of import *quotas* to see if our basic result from the previous paragraph also holds in the case of quotas. Studies of alien species management that incorporate these aspects of the problem into the analysis will provide additional insights into a management problem that has considerable economic and ecological implications.

References

Barbier, E.B., and Shogren, J.F. 2004. Growth with endogenous risk of biological invasion, *Economic Inquiry*, 42, 587-601.

Batabyal, A.A. 2004. A research agenda for the study of the regulation of invasive species introduced unintentionally via maritime trade, *Journal of Economic Research*, 9, 191-216.

Batabyal, A.A., and Beladi, H. 2001. Aspects of the theory of financial risk management for natural disasters, *Applied Mathematics Letters*, 14, 875-880.

Batabyal, A.A., and Beladi, H. 2006. International trade and biological invasions: A queuing theoretic analysis of the prevention problem, *European Journal of Operational Research*, 170, 758-770.

Batabyal, A.A., Beladi, H., and Koo, W.W. 2005. Maritime trade, biological invasions, and the properties of alternate inspection regimes, *Stochastic Environmental Research and Risk Assessment*, 19, 184-190.

Batabyal, A.A., and Lee, D.M. 2006. The infinitesimal, the deterministic, and the probabilistic: Alternate container inspection policies in invasive species management, *Biological Invasions*, 8, 1663-1671.

Batabyal, A.A., and Nijkamp, P. 2005. On container versus time based inspection policies in invasive species management, *Stochastic Environmental Research and Risk Assessment*, 19, 340-347.

Batabyal, A.A., and Nijkamp, P. 2007. The stochastic arrival of alien species and the number of and the damage from biological invasions, *Ecological Economics*, 62, 277-280.

Batabyal, A.A., and Nijkamp, P. 2009. Two aspects of waste management from the viewpoints of a waste generator and a recipient, *Applied Economics Letters*, 16, 337-341.

Biswas, A.K., and Marjit, S. 2007. Preferential trade and mis-invoicing: Some analytical implications, *International Review of Economics and Finance*, 16, 130-138.

Brander, J.A., and Spencer, B. 1984. Tariff protection and imperfect competition, in H. Kierzkowski, (Ed.), *Monopolistic Competition and International Trade*, Oxford University Press, Oxford, UK.

Costello, C., and McAusland, C. 2003. Protectionism, trade, and measures of damage from exotic species introductions, *American Journal of Agricultural Economics*, 85, 964-975.

Cox, G.W. 1993. *Conservation Ecology*. W.C. Brown Publishers, Dubuque, IA.

Eiswerth, M.E., and Johnson, W.S. 2002. Managing nonindigenous invasive species: Insights from dynamic analysis, *Environmental and Resource Economics*, 23, 319-342.

Feenstra, R.C. 2004. *Advanced International Trade*. Princeton University Press, Princeton, NJ.

Helpman, E., and Krugman, P.R. 1989. *Trade Policy and Market Structure*. MIT Press, Cambridge, MA.

Horan, R.D., Perrings, C., Lupi, F., and Bulte, E.H. 2002. Biological pollution prevention strategies under ignorance: the case of invasive species, *American Journal of Agricultural Economics*, 84, 1303-1310.

Horstmann, I., and Markusen, J.R. 1986. Up the average cost curve: Inefficient entry and the new protectionism, *Journal of International Economics*, 20, 225-247.

Jenkins, P. 1996. Free trade and exotic species introductions, *Conservation Biology*, 10, 300-302.

Kulkarni, V.G. 1995. *Modeling and Analysis of Stochastic Systems*. Chapman and Hall, London, UK.

Margolis, M., Shogren, J., and Fischer, C. 2005. How trade politics affect invasive species control, *Ecological Economics*, 52, 305-313.

McAusland, C., and Costello, C. 2004. Avoiding invasives: Trade-related policies for controlling unintentional exotic species introductions, *Journal of Environmental Economics and Management*, 48, 954-977.

Nunes, P.A.L.D., and Van den Bergh, J.C.J.M. 2004. Can people value protection against invasive marine species? Evidence from a joint TC-CV survey in the Netherlands, *Environmental and Resource Economics*, 28, 517-532.

Office of Technology Assessment (OTA). 1993. *Harmful Non-indigenous Species in the United States*, OTA-F-565. Government Printing Office, Washington, DC.

Olson, L.J., and Roy, S. 2002. The economics of controlling a stochastic biological invasion, *American Journal of Agricultural Economics*, 84, 1311-1316.

Parai, A.K. 1999. Profit tax and tariff under international oligopoly, *International Review of Economics and Finance*, 8, 317-326.

Perrings, C., Williamson, M., and Dalmazzone, S., 2000. Introduction, in C. Perrings, M. Williamson, and S. Dalmazzone, (Eds.), *The Economics of Biological Invasions*. Edward Elgar, Cheltenham, UK.

Pimentel, D., Lach, L., Zuniga, R., and Morrison, D. 2000. Environmental and economic costs of nonindigenous species in the United States, *BioScience*, 50, 53-65.

Prestemon, J.P., Zhu, S., Turner, J.A., Buongiorno, J., and Li, R. 2006. Forest product trade impacts of an invasive species: Modeling structure and international trade-offs, *Agricultural and Resource Economics Review*, 35, 128-143.

Ross, S.M. 1996. *Stochastic Processes*, 2nd edition. Wiley, New York, NY.

Simberloff, D., Schmitz, D., and Brown, T. (Eds.). 1997. *Strangers in Paradise*. Island Press, Washington, DC.

Tijms, H.C. 2003. *A First Course in Stochastic Models*. Wiley, Chichester, UK.

USDA, APHIS, and USFS. 2000. *Pest Risk Assessment for Importation of Solid Wood Packing Materials into the United States*. http://www.aphis.usda.gov/ppq/pra/swpm/complete.pdf

Vishwasrao, S., Gupta, S., and Benchekroun, H. 2007. Optimum tariffs and patent length in a model of north-south technology transfer, *International Review of Economics and Finance*, 16, 1-14.

Vitousek, P.M., D'Antonio, C.M., Loope, L.L., and Westbrooks, R. 1996. Biological invasions as global environmental change, *American Scientist*, 84, 468-478.

Yang, Z., and Perakis, A.N. 2004. Multiattribute decision analysis of mandatory ballast water treatment measures in the US great lakes, *Transportation Research Part D*, 9, 81-86.

Zhao, Z., Wahl, T.I., and Marsh, T.L. 2006. Invasive species management: Foot-and-mouth disease in the U.S. beef industry, *Agricultural and Resource Economics Review*, 35, 98-115.

CHAPTER 6

MARITIME TRADE, BIOLOGICAL INVASIONS, AND THE PROPERTIES OF ALTERNATE INSPECTION REGIMES - WITH H. BELADI AND W.W. KOO

Abstract. We analyze the problem of preventing biological invasions caused by ships transporting internationally traded goods between countries and continents. Specifically, we ask the following question: Should a port manager have a small number of inspectors inspect arriving ships less stringently or should this manager have a large number of inspectors inspect the same ships more stringently? We use a simple queuing-theoretic framework and show that if decreasing the economic cost of regulation is very important then it makes more sense for the port manager to choose the less stringent inspection regime. In contrast, if reducing the damage from biological invasions is more salient, then the port manager ought to pick the more stringent inspection regime.

1. Introduction

Maritime trade accounts for a significant proportion of total international trade in the world. Ships are the key vehicle in maritime trade and, today, ships are commonly used to transport a whole host of goods between different countries. It is certainly true that international trade in goods is beneficial to the nations involved in such trade. This notwithstanding, as Heywood (1995) and Parker *et al.* (1999) have pointed out, in addition to transporting goods between countries, ships have also managed to transport—in their ballast water—a variety of alien plant and animal species from one geographical region to another.[32] These alien species have frequently succeeded in invading their new habitats and the ensuing biological invasions have turned out to be extremely costly to the nations in which these novel habitats are located. For the United States alone, the dollar value of these costs is staggering. Here are two examples: First, the Office of Technology Assessment (OTA (1993)) has estimated that the Russian wheat aphid caused $600 million worth of crop damage between 1987 and 1989. Second and more generally, Pimentel *et al.* (2000) have approximated the total costs of all non-native species at around $137 billion per year.

32As Batabyal and Beladi (2006) have noted, the primary method of marine alien species introduction is through the dumping of ballast water. Cargo ships generally carry ballast water in order to increase vessel stability when they are not carrying full loads. When these ships come into port, this ballast water must be discharged before cargo can be loaded. It is estimated that over 4000 species of invertebrates, algae, and fishes are being moved around the world in ship ballast tanks every day. Consider the case of Canada. It has been estimated that as much as 13 billion gallons or 50 million metric tonnes of overseas ballast water enters Canadian coastal ports every year. A recent analysis by the Smithsonian Environmental Research Center (SERC) in Edgewater, Maryland computed that a liter of ballast water generally contains several billion organisms similar to viruses and up to 800 million bacteria. Two web sites that provide useful information on these issues are http://www.fundyforum.com/profile_archives and www.serc.si.edu

As noted by Perrault and Muffett (2002) and Settle and Shogren (2004), invasive species give rise to economic costs and to biological damage. For instance, Vitousek *et al.* (1996) have shown that alien species can change ecosystem processes, act as vectors of diseases, and decrease biological diversity. Further, Cox (1993) has pointed out that out of 256 vertebrate extinctions with a known cause, 109 are the result of biological invasions. In other words, biological invasions have often been a great threat to society.

Although social scientists have, very recently, recognized the salience of the problem of biological invasions, it is still true that "the economics of the problem has...attracted little attention" (Perrings *et al.* (2000, p. 11)). Consequently, our knowledge of the economics of biological invasions in general and the regulation of biological invasions in particular is very incomplete. Now, from a regulatory standpoint, there are several actions that a regulator can take to grapple with the problem of biological invasions. Following Batabyal and Beladi (2006), it is helpful to separate these actions into pre-invasion and post-invasion actions. The point of taking pre-invasion actions is to prevent alien species from invading a novel habitat. Therefore, the reader should think of *pre-invasion* actions as essentially *preventive* in nature. In contrast, post-invasion actions involve the optimal control of an alien species, given that this species has already invaded a novel habitat.

The small extant literature on the economics of biological invasions has, for the most part, addressed the desirability of actions in the *post-invasion* scenario. Here are four examples of such studies. First, Barbier (2001) has shown that the economic effect of a biological invasion can be ascertained by analyzing the nature of the interaction between the alien and the native species. He points out that the economic effect depends on whether this interaction involves interspecific competition or dispersion. Second, Eiswerth and Johnson (2002) have analyzed an intertemporal model of alien species stock management. They demonstrate that the optimal level of management effort is responsive to ecological factors that are not only species and site specific but also random. Third, Olson and Roy (2002) have used a probabilistic framework to investigate the conditions under which it is optimal to wipe out the alien species and conditions under which it is not optimal to do so. Finally, Eiswerth and van Kooten (2002) have shown that, in some situations, it is possible to use information furnished by experts to develop a model in which it is optimal to not decimate but instead control the spread of an invasive species.

To the best of our knowledge, only two papers have theoretically studied the prevention problem, that is, the regulation of a potentially damaging alien species before invasion. These two papers are Horan *et al.* (2002) and Batabyal and Beladi (2006). There are two key differences between our chapter and Horan *et al.* (2002). First, we do not compare the attributes of management strategies under full information and under uncertainty. Second, we use a simple queuing-theoretic framework to shed light on the properties of two inspection regimes that embody different ideas about the economic cost of regulatory activities.

Batabyal and Beladi (2006) is the paper that is most closely related to the present chapter. This paper studies optimization problems arising from the steady state analysis of two multi-person inspection regimes. Although this paper does say something about alternate inspection regimes, it does not address the following cental question that we analyze in the present chapter: When attempting to prevent a biological invasion by inspecting the ballast water of ships, is it a better idea for a port manager to have a *small* number of inspectors inspect arriving ships *less* stringently or should this manager have a *large* number of inspectors inspect the same arriving ships *more* stringently? We use a simple queuing-theoretic framework to show that the answer to the above question depends on the port manager's criterion function. In particular, we focus on two different criteria and show that in one case it makes more sense to have a small number of inspectors inspect arriving ships less stringently and in the other case it is more advantageous to have a large number of inspectors inspect arriving ships more stringently. The reader should note that the basic question that we are addressing in this chapter is entirely consistent with the agenda for research on the regulation of invasive species recently proposed by Batabyal (2004).

The rest of this chapter is organized as follows. Section 2 first provides a primer on queuing theory and then it describes the two queuing-theoretic models that we use to analyze the choice of inspection regime question facing our port manager. Section 3 first focuses on the "average wait of a ship in the port system" or *AWS* criterion for our port manager and then this section analyzes the relative desirability of the two queuing-theoretic inspection regimes described in section 2. Section 4 first focuses on the "average wait of a ship in queue" or *AWQ* criterion for our port manager and then this section undertakes the same analysis as in section 3. Section 5 concludes and offers suggestions for future research on the subject of this chapter.

2. Queuing Theory and the Choice of Inspection Regime

2.1. A primer on queuing theory

The purpose of queuing theory is to mathematically analyze waiting lines or queues.[33] At an elementary level, all queuing models have three attributes. Specifically, they can be delineated by (i) a random

33 Wolff (1989), Taylor and Karlin (1998), and Ross (2003) contain fine textbook accounts of queuing theory.

arrival process, (ii) a probabilistic service time distribution function, and (iii) the deterministic number of available servers. In the queuing models of this chapter, the arrival process is the Poisson process. In this case, the times between successive arrivals are exponentially distributed and the exponential distribution is memoryless. Therefore, the Poisson arrival process is routinely described by the letter *M*.

The service times are obviously random and hence these times can, in principle, be arbitrarily distributed. However, in the queuing models of this chapter, these services times are exponentially distributed and hence they are also memoryless. Hence, once again, the letter *M* is used to symbolize the service time distribution functions. Finally, the non-stochastic number of servers is typically denoted by some positive integer.

Now, consistent with the approach taken in Batabyal (2004) and Batabyal and Beladi (2006), the first inspection regime that we analyze corresponds to the *M*/*M*/1 queuing model and the second inspection regime that we study corresponds to the *M*/*M*/2 queuing model. The reader should note that, in both these tripartite representations, we are referring to inspection regimes in which the arrival process of ships is

Poisson, the time it takes to inspect a ship is exponentially distributed, and the number of inspectors equals either one or two.[34]

2.2. A stylized depiction of biological invasions

Consider a stylized, publically owned port in a particular coastal region of some country. Ships with ballast water arrive at this port to load cargo and to then carry this cargo to some other port. On occasion, ships that come into our port with ballast water will first unload cargo and then load new cargo for shipment to some other port in the world. In either case, the arrival of these ships coincides with the arrival of a whole host of possibly injurious biological organisms. We assume that the arrival rate of these biological organisms is proportional to the arrival rate of the ships. As such, we shall not model these biological organisms directly. Instead, we shall focus on the ships that bring these organisms to our port by means of their ballast water. The arrival process of the ships in our port represents the arrival process for the queuing-theoretic inspection regimes that we analyze in this chapter. Now, consistent with the discussion in the last paragraph of section 2.1, we assume that the ships in question arrive at our port in accordance with a Poisson process with rate β.

34It should be noted that different modeling techniques involve different cross space and time variation features for the underlying physical processes and the associated uncertainties.

The manager of our port is interested in precluding invasions by the possibly injurious biological organisms. Therefore, arriving ships must be inspected before they can either load or unload cargo. In the first model that we analyze, our port has a single inspector and in the second model that we focus on, our port has two inspectors. This means that at any particular point in time, our port will be able to simultaneously inspect either one or two ships. Ships are inspected on a first-come-first-served basis. If more than one or two ships arrive at our port during a particular time interval then the ships that are not already being inspected must wait in queue. The reader should think of this description as follows: Our port has either one or two docks and one inspector is assigned to a dock. Hence, at any particular point in time, a maximum of either one or two ships can be docked and inspected. The port system consists of ships that are being inspected, ships that are waiting in queue, the one or two inspectors, and the port manager.

Given that we are studying the prevention of biological invasions, ideally, our port manager would like the inspectors to have a zero tolerance policy. However, such a policy may be very costly to implement. As such, we suppose that our port manager is confronted with a choice between two scenarios. In the first scenario, the manager has one inspector inspect the ballast water of arriving ships not very stringently. In the second scenario, this manager has two inspectors inspect the ballast water of arriving ships more stringently. Examples of activities that an inspector might undertake include (i) the shipboard filtration of ballast water, (ii) the treatment of ballast water with heat, chemicals, and ultraviolet radiation, and (iii) the shore based treatment of ballast water.

Inspections generally require varying amounts of time. For example, if an inspector knows that a particular ship has taken on ballast water in an area where there are no known biological invaders, then he may be able to clear a ship relatively speedily. In contrast, if a specific ship has taken on ballast water during a phytoplankton bloom, then the probability that this ship's ballast water will contain injurious organisms is much higher, and hence more time will be required to clear this ship. The upshot of this discussion is that the inspection times—and hence the stringency of inspections—are *random* variables. Given this state of affairs, we now make the following two assumptions regarding the stringency of inspections. In the first scenario of the previous paragraph, our single inspector inspects arriving ships at rate 2γ where γ is a positive constant. In the second scenario of the previous paragraph, the two inspectors inspect arriving ships at rate γ. A faster rate of inspection implies lower stringency and a slower rate of inspection implies higher stringency. Therefore, the inspection regime of the first scenario is *less* stringent than the inspection regime of the second scenario because $2\gamma > \gamma$.

We now have all the necessary parts of our two queuing-theoretic inspection regimes. Using the language of queuing theory, the first or less stringent inspection regime is a *M/M/*1 model with inspection rate 2γ and the second or more stringent inspection regime is a *M/M/*2 model with inspection rate γ. The reader should note the way in which we have mathematically characterized the cental question of this chapter: When attempting to prevent a biological invasion by inspecting the ballast water of ships, is it a better idea for a port manager to have a single inspector inspecting arriving ships at rate 2γ or should this manager have two inspectors inspecting the same ships at rate γ? We now proceed to the theoretical discussion of the two queuing-theoretic inspection regimes for the case in which our port manager focuses on the "average wait of a ship in the port system" or *AWS* criterion.

3. The *AWS* Criterion

Inspection activities that result in the prevention of a biological invasion by alien plant or animal species clearly result in benefits to the citizens of the coastal region under consideration. However, during the time that arriving ships are being inspected, there is in particular no loading or unloading of cargo, and hence in general, economic activity resulting from maritime trade is at a standstill. This temporary stoppage of economic activities imposes costs on the economy of our coastal region and one way to proxy this cost is to suppose that this cost is directly proportional to the average wait of a ship in the port system. In this way of looking at the problem, the longer (shorter) this average wait in the port system or *AWS,* the larger (smaller) the costs from the suspension of economic activities. Therefore, in having inspectors inspect arriving ships for potentially injurious biological organisms, our port manager will want to keep *AWS* as low as possible. Now, in the rest of this section, we suppose that our port manager has this *AWS* (proxy for economic cost) criterion in mind when he is choosing between the less stringent and the more stringent inspection regimes.

3.1. M/M/2 inspection regime

Let us now calculate *AWS* for the *M/M/*2 inspection regime with inspection rate γ. We proceed by means of three steps. First, we explain the notion of a stationary probability. To this end, let $Z(t)$ denote the total number of ships in our port at time t. Then, we define P_n, $n \geq 0$ to be

$$P_n = \lim_{t \to \infty} Prob\{Z(t)=n\}. \quad (1)$$

In words, P_n is the long run or stationary probability that there are exactly n ships in our port system.

The second step requires us to set up and solve the so called balance equations. Now, following the procedure described in Ross (2003, pp. 480-483), the specific balance equations we seek are

$$\beta P_0 = \gamma P_1, \tag{2}$$

$$(\beta+\gamma)\ P_1 = \beta P_0 + 2\gamma P_2, \tag{3}$$

and

$$(\beta+2\gamma) P_n = \beta P_{n-1} + 2\gamma P_{n+1},\ n \geq 2. \tag{4}$$

Let $\psi=\beta/\gamma$. Then, following the above cited procedure, it can be shown that the solutions to the balance equations (2)-(4) are $P_n=\psi^n/2^{n-1}P_0$. In words, the stationary or long run or limiting probability that there are n ships in our port system is a multiplicative function of the stationary probability that there are zero ships in this port system. Now, it should be clear to the reader that $\sum_{n=0}^{n=\infty} P_n=1$. This last condition tells us that

$$P_0 = \frac{1-\psi/2}{1+\psi/2} = \frac{2-\psi}{2+\psi}. \tag{5}$$

Note that equation (5) above only makes sense when $2 \geq \psi$. Because $\psi=\beta/\gamma$, this means that we must have $2\gamma \geq \beta$.

Third, now that we know the P_n, we can calculate the average number of ships in our port system S. From Ross (2003, p. 483), we infer that $S=\sum_{n=0}^{n=\infty} nP_n$. Using this expression and then simplifying, we get

$$S=\frac{4\beta\gamma}{(2\gamma+\beta)(2\gamma-\beta)}. \tag{6}$$

Now, having determined the expected number of ships in our port system, we can ascertain the average wait of a ship in our port system or *AWS* by using a well known result in queuing theory. This result—see Taylor and Karlin (1998, p. 542)—tells us that $AWS=S/\beta$.[35] Using this last finding, we get

$$AWS_{M/M/2}=\frac{4\gamma}{(2\gamma+\beta)(2\gamma-\beta)}. \tag{7}$$

This completes our three step derivation of the *AWS* criterion for the *M/M/2* inspection regime. We now derive the same criterion for the *M/M/1* inspection regime.

35This result is important because it relates two important metrics in queuing models, i.e., the average queue size and the average customer waiting time in the stationary state. In the context of our chapter, the average queue size is the average number of ships in our port system and the average customer waiting time is our *AWS* criterion.

3.2. M/M/1 inspection regime

The derivation of the *AWS* criterion for the *M/M/*1 inspection regime is considerably simpler. In particular, to obtain the criterion we seek, we shall modify a result stated in Taylor and Karlin (1998, p. 551). This gives us

$$AWS_{M/M/1}=\frac{1}{(2\gamma-\beta)}. \tag{8}$$

Given equations (7) and (8), we now proceed to discuss the central question of this chapter for the *AWS* criterion.

3.3. Discussion

When seeking to prevent a biological invasion by inspecting the ballast water of ships, is it a better idea for a port manager to have a *small* number of inspectors inspect arriving ships *less* stringently or should this manager have a *large* number of inspectors inspect the same arriving ships *more* stringently? Our model of a small number of inspectors inspecting arriving ships less stringently is the *M/M/*1 regime with inspection rate 2γ. Similarly, our model of a large number of inspectors inspecting arriving ships more stringently is the *M/M/*2 regime with inspection rate γ. Consequently, we are now in a position to answer the above question for the *AWS* criterion.

Let us suppose that for the *M/M/*1 inspection regime, the condition $\beta<2\gamma$ holds. That is, the denominator of the right-hand-side (RHS) of equation (8) is positive. In words, this simply means that our queuing-theoretic inspection regime is stable. Now, $\beta<2\gamma$ implies that $\beta+2\gamma<4\gamma$ and this inequality tells us that $1<\{4\gamma/(\beta+2\gamma)\}$. This last inequality and equations (7) and (8) together tell us that $AWS_{M/M/1} < AWS_{M/M/2}$. Hence, we can now state

PROPOSITION 1. *AWS* is lower when our port manager chooses a small number of inspectors to inspect arriving ships less stringently.

Note that if one finds the queue empty in the *M/M/*2 inspection regime then it would not be beneficial to have two inspectors. Instead, one would always be better off with one faster (less stringent) inspector. This is the intuitive explanation for the Proposition 1 result that *AWS* is lower when our port manager selects a small number of inspectors to inspect arriving ships less stringently. In addition, the reader should note that the result contained in Proposition 1 is not based on general distribution functions for either the interarrival times or the inspection times. Instead, this result is based on Markovian distribution functions for the interarrival and the inspection times. A justification for the use of these distribution functions has been given in section 2.1.

Given the significance of the problem of biological invasions, it is perhaps intuitively plausible that our port manager ought to choose the inspection regime involving more stringent inspections and a larger number of inspectors, i.e., the *M/M/*2 inspection regime with inspection rate γ. However, using a larger number of more stringent inspectors imposes costs in the sense that the *M/M/*2 regime stops all economic activity in our port for a relatively long period of time. We modeled these costs by supposing that our port manager wishes to make the *AWS* criterion as small as possible. Our analysis thus far shows that there is a tension between *AWS* minimization and inspection stringency and that tension is this: If greater inspection stringency with a large number of inspectors is desired, then economic costs measured by the *AWS* criterion will be relatively high. Conversely, if our port manager is willing to live with less stringent inspections with a smaller number of inspectors, then *AWS* will be relatively low. We now consider the choice of inspection regime question for an alternate criterion function for our port manager.

4. The *AWQ* Criterion

We have already noted that the temporary stoppage of economic activities in our port imposes costs on the economy of our coastal region. However, the *AWS* criterion is not the only criterion that we can use in thinking about these costs. Suppose we adopt a somewhat looser interpretation of these costs and say that the loading and unloading of cargo may proceed on a ship that is currently being inspected but that such activities may not take place on ships that have yet to be inspected, i.e., those that are still in queue. In this way of looking at the problem, only ships that are in *queue* impose economic costs. We shall refer to this looser criterion as the "average wait of a ship in queue" or *AWQ* criterion. The reader will note that in this looser interpretation of costs, once inspection on a particular ship has commenced, there is a presumption of innocence rather than guilt. The basic question before us now is this: Does the section 3 answer to our port manager's inspection regime choice question change when we adopt the *AWS* criterion? We now answer this question.

4.1. M/M/2 inspection regime

To compute *AWQ* for the *M/M/2* inspection regime with rate γ, we begin by noting that $AWQ_{M/M/2}=AWS_{M/M/2}-1/\gamma$. Using this expression along with equation (7), it is easy to see that

$$AWQ_{M/M/2}=\frac{\beta^2}{\gamma(2\gamma+\beta)(2\gamma-\beta)}. \tag{9}$$

We now derive *AWQ* for the *M/M/1* inspection regime with inspection rate 2γ.

4.2. M/M/1 inspection regime

In this case, the corresponding relationship between $AWQ_{M/M/1}$ and $AWS_{M/M/1}$ is that $AWQ_{M/M/1}=AWS_{M/M/1}-1/2\gamma$. Using this expression and equation (8) together, we conclude that

$$AWQ_{M/M/1}=\frac{\beta}{2\gamma(2\gamma-\beta)}. \tag{10}$$

Given equations (9) and (10), we are now in a position to discuss the central question of this chapter for the *AWQ* criterion.

4.3. Discussion

To determine whether the section 3 answer to the choice of inspection regime question changes, we have to compare the RHSs of equations (9) and (10). Doing this, it is clear that

$$AWQ_{M/M/1} > AWQ_{M/M/2} \Leftrightarrow \frac{1}{2} > \frac{\beta}{2\gamma+\beta} \Leftrightarrow 2\gamma > \beta. \tag{11}$$

We now invoke the section 3.3 assumption about the stability of the *M/M/*1 inspection regime. Recall that this stability assumption is that the condition $2\gamma > \beta$ holds. Using this stability assumption along with equation (11), it is clear that $AWQ_{M/M/1} > AWQ_{M/M/2}$. Therefore, we can now state

PROPOSITION 2. *AWQ* is lower when our port manager selects a large number of inspectors to inspect arriving ships more stringently.

Given the salience of the problem of biological invasions, it makes intuitive sense that our port manager ought to select the inspection regime with more stringent inspections and a larger number of inspectors, i.e., the *M/M/2* regime with inspection rate γ. Proposition 2 tells us that this intuitive line of reasoning is correct when the port manager focuses on the *AWQ* criterion. Proposition 2 also tells us that the section 3 answer to the choice of inspection regime question *does change* when our port manager focuses on the *AWQ* criterion. This means that whether it is better to have a small number of inspectors inspect arriving ships less stringently or to have a large number of inspectors inspect the same arriving ships more stringently depends on which economic cost criterion our port manager adopts.

More generally, there is a tension between economic cost reduction on the one hand and biological damage control on the other. If total economic cost reduction is very important, then the port manager will focus on the stronger *AWS* criterion and this focus means that the chosen inspection regime will be relatively less stringent. In contrast, if biological damage control is very important, then our port manager will select the more stringent *M/M/2* regime and, as we have just seen, this selection is optimal only when the manager concentrates on the weaker AWQ criterion.

5. Conclusions

Maritime trade in goods by means of ships often results in injurious invasions of new habitats by alien plant and animal species. Consequently, if an appropriate authority, such as a port manager's objective is to prevent biological invasions, then this manager must inspect arriving ships for potentially deleterious biological organisms. Given this state of affairs, what kind of inspection regime should this manager have in place? In particular, is it a better idea for this manager to have a *small* number of inspectors inspect arriving ships *less* stringently or should this manager have a *large* number of inspectors inspect the same ships *more* stringently? Our analysis shows that if decreasing economic cost is significant, then it makes more sense for the port manager to choose the inspection regime with fewer inspectors and less stringent inspections. On the other hand, if reducing the damage from biological invasions is more salient, then the manager ought to pick the inspection regime with more inspectors and more stringent inspections.

The analysis in this chapter can be extended in a number of directions. In what follows, we propose two potential extensions of this chapter's research. First, we analyzed the inspection regime choice question in a very simple manner. Therefore, it would be instructive to study the properties of *M/M/m* and *M/M/n* inspection models where m and n are positive integers, $m<n$, and the inspection rates in the two models are different. Second, it would also be useful to eschew the assumption that the inspection times are exponentially distributed and examine more general *M/G/m* and *M/G/n* inspection models where G denotes a general distribution function. Studies of maritime trade driven biological invasions that incorporate these aspects of the inspection regime choice question into the analysis will provide additional insights into a problem that has considerable economic and biological ramifications.

References

Barbier, E.B. 2001. A note on the economics of biological invasions, *Ecological Economics*, 39, 197-202.

Batabyal, A.A. 2004. A research agenda for the study of the regulation of invasive species introduced unintentionally via maritime trade, *Journal of Economic Research*, 9, 191-216.

Batabyal, A.A., and Beladi, H. 2006. International trade and biological invasions: A queuing theoretic analysis of the prevention problem, *European Journal of Operational Research*, 170, 758-770.

Cox, G.W. 1993. *Conservation Ecology*. W.C. Brown Publishers, Dubuque, IA.

Eiswerth, M.E., and Johnson, W.S. 2002. Managing nonindigenous invasive species: Insights from dynamic analysis, *Environmental and Resource Economics*, 23, 319-342.

Eiswerth, M.E., and van Kooten, G.C. 2002. Uncertainty, economics, and the spread of an invasive plant species, *American Journal of Agricultural Economics*, 84, 1317-1322.

Heywood, V. Ed. 1995. *Global Biodiversity Assessment*. Cambridge University Press, Cambridge, UK.

Horan, R.D., Perrings, C., Lupi, F., and Bulte, E.H. 2002. Biological pollution prevention strategies under ignorance: The case of invasive species, *American Journal of Agricultural Economics*, 84, 1303-1310.

Office of Technology Assessment (OTA). 1993. *Harmful Non-Indigenous Species in the United States*, OTA-F-565, Washington, DC.

Olson, L.J., and Roy, S. 2002. The economics of controlling a stochastic biological invasion, *American Journal of Agricultural Economics*, 84, 1311-1316.

Parker, I.M., Simberloff, D., Lonsdale, W.M., Goodell, K., Wonham, M., Kareiva, P.M., Williamson, M.H., Von Holle, B., Moyle, P.B., Byers, J.E., and Goldwasser, L. 1999. Impact: Toward a framework for understanding the ecological effects of invaders, *Biological Invasions*, 1, 3-19.

Perrault, A.M., and Muffett, W.C. 2002. Turning off the tap: A strategy to address international aspects of invasive alien species, *Review of European Community and International Environmental Law*, 11, 211-224.

Perrings, C., Williamson, M., and Dalmazzone, S. 2000. Introduction in C. Perrings, M. Williamson, and S. Dalmazzone, (Eds.), *The Economics of Biological Invasions*. Edward Elgar, Cheltenham, UK.

Pimentel, D., Lach, L., Zuniga, R., and Morrison, D. 2000. Environmental and economic costs of nonindigenous species in the United States, *BioScience*, 50, 53-65.

Ross, S.M. 2003. *Introduction to Probability Models*, 8th edition. Academic Press, San Diego, CA.

Settle, C., and Shogren, J.F. 2004. Modeling native-exotic species within Yellowstone lake, *American Journal of Agricultural Economics*, 84, 1323-1328.

Taylor, H.M., and Karlin, S. 1998. *An Introduction to Stochastic Modeling*, 3rd edition. Academic Press, San Diego, CA.

Vitousek, P.M., D'Antonio, C.M., Loope, L.L., and Westbrooks, R. 1996. Biological invasions as global environmental change, *American Scientist*, 84, 468-478.

Wolff, R.W. 1989. *Stochastic Modeling and the Theory of Queues*. Prentice-Hall, Englewood Cliffs, NJ.

CHAPTER 7

ON CONTAINER VERSUS TIME BASED INSPECTION POLICIES IN INVASIVE SPECIES MANAGEMENT - WITH P. NIJKAMP

Abstract. We study the problem of precluding biological invasions caused by ships transporting internationally traded goods in containers between different regions of the world. Using the long run expected net cost (*LRENC*) of inspections as the apposite managerial objective, we address the following important question: Given that inspection is a *cyclical* activity, is the *LRENC* lower when a port manager's inspector inspects cargo upon the arrival of a specified number of containers (container policy) or is this *LRENC* lower when this inspector inspects cargo at fixed points in time (temporal policy)? We construct a queuing theoretic model and show that in an inspection cycle, irrespective of whether the inspection policy choice is made on the basis of an explicit optimization exercise or on the basis of rules of thumb, the container policy is superior to the temporal policy, because the container policy results in lower *LRENC* from inspection activities.

1. Introduction

It is now well known that maritime trade in goods comprises a significant fraction of the world's total international trade in goods.

Ships are the primary vehicle in maritime trade and, in contemporary times, ships are routinely used to transport a whole host of goods in containers from one region of the world to another. Although there are clear gains to the involved parties from such voluntary trade between the different regions of the world, researchers have increasingly noted that the magnitude of these gains is likely to be less than what most observers have hitherto believed. Why is this the case? As Heywood (1995), Parker *et al.* (1999), and Batabyal (2004) have pointed out, this is because, in addition to transporting goods in containers between regions, ships have also succeeded in transporting a variety of alien plant and animal species[36] from one geographical region to another.

Broadly speaking, ships have transported non-native species in two main ways. First, a variety of marine alien species have been introduced into a region, often unwittingly, by ships dumping their ballast water. Cargo ships routinely carry ballast water, in order to increase vessel stability when they are not carrying full loads. When these ships come into port, this ballast water must be released before cargo can be loaded. This method of species introductions is important and very recently the problem of managing invasive species that have

36In the rest of this chapter, we shall use the terms "alien species," "invasive species," and "non-native species" interchangeably.

been introduced into a particular region by means of the dumping of ballast water has received some attention in the literature.[37]

The second way in which alien species have been introduced into a specific region is by means of the containers that ships commonly use to transport cargo from one region to another. Indeed, non-native species can remain undetected in containers for long periods of time. In addition, the material (such as wood) that is used to pack the cargo in the containers may itself contain alien species. In this connection, a joint report from the United States Department of Agriculture (USDA), the Animal and Plant Health Inspection Service (APHIS) and the United States Forest Service (USFS) has noted that approximately 51.8% of maritime shipments contain solid wood packing materials and that infection rates for solid wood packing materials are non-trivial (USDA, APHIS, and USFS (2000, p. 25)). For example, inspections of wooden spools from China revealed infection rates between 22% and 24% and inspections of braces for granite blocks imported into Canada were found to contain live insects 32% of the time (USDA, APHIS, and USFS (2000, pp. 27-28)).

37For additional details on this subject, the reader should consult Yang and Perakis (2004), Batabyal *et al.* (2005), and Batabyal and Beladi (2006), and the many references in these three papers.

The non-native species that we have been discussing generally thus far have frequently succeeded in invading their new habitats and the resulting biological invasions have turned out to be very costly for the regions in which these novel habitats are located. For the United States alone, the dollar value of these costs is mind boggling. To see this, consider the following two examples. First, the Office of Technology Assessment (OTA (1993)) has calculated that the Russian wheat aphid caused $600 million worth of crop damage between 1987 and 1989. Second, Pimentel *et al.* (2000) have estimated the total costs of all non-native species at around $137 billion per year.

In addition to the economic costs that we have just noted, invasive species have given rise to significant biological damage as well. In this regard, Vitousek *et al.* (1996) have demonstrated that non-native species can change ecosystem processes, act as vectors of diseases, and decrease biological diversity. Further, Cox (1993) has noted that out of 256 vertebrate extinctions with a known cause, 109 are the result of biological invasions. The implication of the discussion in this and the preceding paragraph is clear: Biological invasions have frequently been a great menace to society.

Very recently, economists and regional scientists have acknowledged the salience of the problem of biological invasions. Even so, it is still true that "the economics of the problem has...attracted little attention" (Perrings *et al.* (2000, p. 11)). Therefore, our knowledge of the economic and the management aspects of invasive species is incomplete. Now, from the standpoint of a manager, there are various actions that he can take to deal with the problem of biological invasions. Following Batabyal and Beladi (2006), it is helpful to separate these actions into pre-invasion and post-invasion actions. The purpose of pre-invasion actions is to preclude alien species from invading a new region. Therefore, the reader should think of pre-invasion actions as fundamentally prophylactic in nature. In contrast, post-invasion actions involve the optimal control of an alien species, given that this species has already invaded a novel region.

The focus of the small extant literature on biological invasions has, for the most part, been on the desirability of actions in the *post-invasion* scenario. Here are four examples of such analyses. First, Barbier (2001) has shown that the economic impact of a biological invasion can be determined by examining the nature of the interaction between the alien and the native species. He notes that the economic impact depends on whether this interaction involves interspecific competition or dispersion. Second, Eiswerth and Johnson (2002) have studied a dynamic model of alien species stock management. They show that the optimal level of management effort is responsive to ecological factors that are not only species and site specific but also random. Third, Olson and Roy (2002) have used a stochastic framework to explore the conditions under which it is optimal to wipe out an alien species and conditions under which it is not optimal to do so. Finally, Eiswerth and van Kooten (2002) have shown that, in some circumstances, it is possible to use information provided by experts to develop a model in which it is optimal to not eradicate, but instead control, the spread of an invasive species.

Our search of the pertinent literature located three papers that have theoretically analyzed the prevention problem; that is, the regulation of a possibly injurious alien species before it has invaded a particular region. These three papers are Horan *et al.* (2002), Batabyal *et al.* (2005), and Batabyal and Beladi (2006). Horan *et al.* (2002) compare the properties of management strategies under full information and under uncertainty. Batabyal and Beladi (2006) study optimization problems stemming from the steady state analysis of two multi-person inspection regimes. Finally, Batabyal *et al.* (2005) show that, if decreasing the economic cost associated with inspections is significant, then it makes more sense for a port manager to choose the inspection regime with fewer inspectors and less stringent inspections. In contrast, if reducing the damage from biological invasions is more salient, then this manager ought to pick the inspection regime with more inspectors and more stringent inspections.

Like Batabyal *et al.* (2005), we also focus on the inspection aspect of the management problem. However, unlike the Batabyal *et al.* (2005) paper, we study a very different question in this chapter. Specifically, using the long run expected net cost (*LRENC*) of inspections as the apposite managerial objective, we address the following important question: Given the cyclical nature of the inspection function, is the *LRENC* of inspections lower when a port manager follows a policy of inspecting cargo upon the arrival of a specified number of containers (the container policy), or is this *LRENC* lower when this manager inspects cargo at fixed points in time (the temporal policy)? We construct a queuing model—that is different from the model used in Batabyal *et al.* (2005)—and show that, in an inspection cycle, irrespective of whether the inspection policy choice is made on the basis of an explicit optimization exercise or on the basis of rules of thumb, the container policy is superior to the temporal policy because the container policy results in lower *LRENC* from inspection activities.

The theoretical framework of this chapter is adapted from Batabyal *et al.* (2001) and Ross (2003, pp. 515-519) and the rest of this chapter is organized as follows. Section 2 first provides a primer on queuing theory and then this section provides a stylized account of biological invasions in the context of a queuing theoretic model of the inspection policy choice problem faced by a port manager. Section 3 analyzes this choice problem for the case in which the port manager wishes to minimize the *LRENC* from cargo inspections by optimally choosing the number of containers to inspect in an inspection cycle. Section 4 studies a similar model. However, in this section, the port manager minimizes the *LRENC* from cargo inspections by optimally choosing the temporal inspection point in an inspection cycle. Section 5 compares the optimized value of the port manager's *LRENC* from sections 3 and 4 and thereby determines which inspection policy results in lower *LRENC*. Section 6 concludes and offers two suggestions for future research on the subject of invasive species management.

2. Queuing Theory and the Choice of Inspection Policy

2.1. Preliminaries

The purpose of queuing theory—see Wolff (1989), Kulkarni (1995), Taylor and Karlin (1998), and Ross (2003) for textbook accounts—is to mathematically analyze waiting lines or queues. Three features are common to all queuing models. Specifically, they can be described by a stochastic arrival process, a probabilistic service time distribution function, and a fixed number of servers. In the queuing model that we employ in this chapter, the arrival process is the Poisson process. Therefore, the times between successive arrivals are exponentially distributed and the exponential distribution is *memoryless* or Markovian. Hence, the Poisson arrival process is commonly described by the letter *M*.

In general, the service times are random variables. Therefore, these times can, in principle, be arbitrarily distributed and this is assumed in our chapter. Therefore, we use the letter G to denote the *general* service time distribution function. Finally, the deterministic number of servers is typically denoted by some positive integer. In this chapter, servers are inspectors. Moreover, our analysis will be conducted from the perspective of a port manager who employs a so-called representative inspector (hereafter inspector). Therefore, we shall work with a single inspector. In the language of queuing theory, our model corresponds to the well known $M/G/1$ queuing model.

2.2. A stylized account of biological invasions

Consider a stylized, publically owned port in a particular coastal region of some country. Upon arrival at this port, ships unload their containers carrying cargo. The arrival of these containers coincides with the arrival of a whole host of conceivably deleterious biological organisms. Now, before this incoming cargo can be moved to various points in the interior of the country under consideration, the containers must first be *inspected* at the inspection facility in our port. The purpose of this protective activity is to ensure that one or more biological invasions—with potentially adverse consequences for the economy and the ecology of the country under study—do not in fact take place. We suppose that the arrival rate of the various biological organisms is proportional to the arrival rate of the containers at the inspection facility. Hence, we shall not model the biological organisms directly. Instead, we shall focus on the containers that bring these organisms to our port. Further, the arrival process of the containers at the inspection facility in our port represents the arrival process for the queuing model that we analyze in this chapter. Finally, we assume that the containers in question arrive at the inspection facility in accordance with a Poisson process with rate α and, to keep things relatively simple, we also suppose that there is no capacity constraint affecting the functioning of the port under study.

The manager of our port is interested in precluding invasions by the possibly injurious biological organisms. As such, in this chapter, his basic choice problem is to determine which of two possible inspection policies he ought to have in place. Now, from the standpoint of invasive species management, note the following key features of inspections. First, inspections are physically costly to undertake. Second, quite apart from the cost of conducting physical inspections, inspections also impose an economic cost on society. This cost arises from the fact that, while containers are being inspected, there is no unloading or loading of cargo and hence economic activity in our port is very slow, if not at a standstill. Third, properly conducted inspections reduce (and perhaps even eliminate) the likelihood of a biological invasion. In the following sections 3 and 4, we shall explicitly model these three features of inspections.

The reader will note that inspections generally require varying amounts of time. For example, if the inspector knows that a batch of containers awaiting inspection are all from a particular country from which either zero or few invaders have emerged in the past, then he may be able to clear this batch of containers relatively quickly. On the other hand, if the containers awaiting inspection are either from several nations or from a nation with a known history of invasive species problems, then more time will generally be needed to clear the containers in question. The upshot of this discussion is that the inspection times are *random* variables. Given this state of affairs, let I denote the inspection time random variable and let $E[I]$ denote its expectation. The reader will recall that I has a *general* cumulative distribution function. Let us denote this function by $G(\cdot)$. The key pieces of our queuing theoretic model are now in place. Therefore, we now systematically analyze the container policy first in section 3 and then the temporal policy in section 4.

3. The Container Policy

Containers arrive at the inspection facility in accordance with a Poisson process with rate α. The port manager's inspector examines the containers that have arrived at the inspection facility and he continues to do so until all the containers that are present have been inspected. When this busy period ends, the inspector leaves the inspection facility and he returns only when N new containers have arrived and are awaiting inspection. As a result of these inspections, the inspector—and ultimately the port manager—incurs costs and obtains benefits from two sources. The first source of net cost—this is the difference between total cost and total benefit—arises from things like the expense of paying the inspector and operating the inspection equipment (a cost) and from the reduction in the likelihood of a biological invasion (a benefit). We model this first source of net cost by supposing that our inspector incurs net cost at the rate of c dollars per container per unit time. The second source of net cost stems from things like the deleterious impact on society from the slowdown in economic activity while the containers are being inspected (a cost) and from the determination of whether the containers actually contain what they are supposed to contain (a benefit). We account for this second source of net cost by supposing that the inspector incurs a net cost of C dollars each time he returns to the inspection facility.

The reader should note that the inspection function is *cyclical* in nature. In other words, when containers have arrived at the facility in our port, the inspector is busy inspecting these containers until there are no more containers waiting to be inspected. This is the *busy* period. Then comes an *idle* period in which the inspector has no specific duties to perform. Then, when N additional containers arrive at the facility to be inspected, a new busy period commences. This busy period is followed by an idle period, and so on and so forth. Given this state of affairs, if we say that a cycle is completed whenever the inspector *returns* to the inspection facility, then the delineation of events in the previous paragraph constitutes a renewal-reward process.[38] Accordingly, we can use the renewal-reward theorem to compute our port manager's *LRENC* from inspection activities.[39] Let *E*[*net cost per cycle*] denote the expected net cost of inspections per cycle. Similarly, let *E*[*length of cycle*] denote the expected length of an inspection cycle. Then, the renewal-reward theorem tells us that the inspector's *LRENC* is given by a specific ratio and that ratio is

$$LRENC = \frac{E[net\ cost\ per\ cycle]}{E[length\ of\ cycle]} \tag{1}$$

Given this setup, our port manager's problem is to choose N optimally to minimize the *LRENC* of inspections.

38See Ross (1996, pp. 132-140) and Ross (2003, pp. 416-425) for more on renewal-reward processes in general and the renewal-reward theorem in particular.

39We shall soon describe the port manager's optimization problem as a long run expected net cost minimization problem. This notwithstanding, the reader should note that, without any substantive changes, we could also have delineated this manager's optimization problem as a long run expected net benefit maximization problem.

Let us now calculate the two expectations on the right hand side (RHS) of equation (1). Consider a time interval of length L_j, which begins at the first time during an inspection cycle in which there are j containers in the facility and ends at the first time thereafter that there are only j-1 containers. Then $\sum_{j=1}^{j=N} L_j$ is the total amount of time that the inspector is *busy* checking containers during an inspection cycle. To this time we now add the average time for which our inspector is *idle* until N new containers arrive at the inspection facility. Doing this gives us

$$E[length\ of\ cycle]=\sum_{j=1}^{j=N} E[L_j]+\frac{N}{\alpha}. \tag{2}$$

Now consider the moment in time when an inspection is about to begin and there are j-1 containers waiting to be inspected. We assume that the inspection times do *not* depend on the order in which the containers are inspected. As such, suppose that the order of inspection is last in, first out or LIFO. Then, as noted in Ross (2003, pp. 322-324), this means that the amount of time it takes to go from j containers waiting in the facility to be inspected to j-1 containers waiting to be inspected has the same distribution as the length of the busy period B of the *M/G/*1 queuing model. Now, from equation 8.31 in Ross (2002, p. 253), it follows that

$$E[L_j]=E[B]=\frac{E[I]}{1-\alpha E[I]}, \tag{3}$$

where *I,* the reader will recall, is the inspection time random variable. For equation (3)—and indeed many of the subsequent equations in this chapter—to make sense, we must have $\alpha E[I]<1$. Therefore, in the rest of this chapter we assume that this inequality holds. The reader should think of this inequality as a stability condition. Specifically, the inequality $\alpha E[I]<1$ gives us a condition for the finiteness of the expected length of the busy period *B* of the *M/G/1* queuing model. Now, using equation (3) to simplify equation (2), we get

$$E[length\ of\ cycle]=\sum_{j=1}^{j=N}\frac{E[I]}{1-\alpha E[I]}+\frac{N}{\alpha}=\frac{N}{\alpha(1-\alpha E[I])}. \quad (4)$$

This completes the task of calculating the expected length of an inspection cycle. We now compute the numerator on the RHS of equation (1) or the expected net cost per inspection cycle incurred by our port manager's inspector.

To compute *E*[*net cost per cycle*], let $\widehat{C_j}$ be the net cost incurred by the inspector during a time period of length L_j, where L_j is as described in the paragraph immediately preceding equation (2). Then, it follows that the total net cost incurred during the busy period of an inspection cycle is $\sum_{j=1}^{j=N}\widehat{C_j}+C$. To this, we now have to add the net cost incurred during the idle period of the inspection cycle. Note that because the containers arrive at the inspection facility in accordance with a Poisson process with rate α, there will be *j* containers in the facility for an exponential amount of time with rate α and the index j runs from 1 to N-1.[40] Therefore, the total expected net cost during the idle period is $c(1+2+3+\ldots+N-1)/\alpha = cN(N-1)/2\alpha$ With this computation in place, the total expected net cost in an inspection cycle or E[*net cost per cycle*] is

$$E[\textit{net cost per cycle}] = \sum_{j=1}^{j=N} E[\widehat{C}_j] + C + \frac{cN(N-1)}{2\alpha}. \tag{5}$$

Our next task is to find an expression for $E[\widehat{C}_j]$. To do this, recall the time interval of length L_j. During this time interval, let S_j be the *sum* of the initial inspection time and the sum of all the times spent in the inspection facility by the containers that have arrived and have been inspected until the L_j time interval ends and there are only j-1 containers in the inspection facility. Mathematically, we can write this as $\widehat{C}_j=(j-1)cL_j+cS_j$. To proceed further, it is important to recognize that S_j is distributed as the *sum* of the times spent in the inspection facility by all the arriving containers during the busy period of the *M/G/*1 queuing model. Let us denote this sum random variable and its expectation by B_S and by $E[B_S]$ respectively. Then, using equation (3) and the expression for $\widehat{C}_j$ above, we get

40 This result follows from proposition 2.2.1 in Ross (1996, p. 64) and proposition 7.2.1 in Ross (2002, p. 203).

$$E[\widehat{C}_j]=(j\text{-}1)cE[L_j]+cE[S_j]=(j\text{-}1)c\frac{E[I]}{1-\alpha E[I]}+cE[B_S]. \tag{6}$$

From Ross (2003, p. 518), we conclude that $E[B_S]=\alpha E[I^2]/\{2(1-\alpha E[I])^2\}+E[I]/(1-\alpha E[I])$. Using this last expression for $E[B_S]$ and equation (6), we can simplify equation (5). This simplification gives

$$E[\textit{net cost per cycle}]=cN[\frac{\alpha E[I^2]}{2(1-\alpha E[I])^2}+\frac{E[I]}{1-\alpha E[I]}]+C+\frac{cN(N-1)}{2\alpha(1-\alpha E[I])}. \tag{7}$$

Now, dividing the RHS of equation (7) by the RHS of equation (4), we get our required expression for the *LRENC* of inspections. That expression is

$$LRENC=\alpha cE[I]+\frac{\alpha^2 cE[I^2]}{2(1-\alpha E[I])}+\frac{c(N-1)}{2}+\frac{\alpha C(1-\alpha E[I])}{N}. \tag{8}$$

Having computed the expression for the *LRENC* of inspections, we are now in a position to state our port manager's optimization problem. Specifically, this manager chooses the number of containers N to minimize the *LRENC* from inspection activities. Formally, our port manager solves

$$\min_{\{N\}}\{\alpha cE[I]+\frac{\alpha^2 cE\left[I^2\right]}{2\left(1-\alpha E\left[I\right]\right)}+\frac{c(N-1)}{2}+\frac{\alpha C\left(1-\alpha E\left[I\right]\right)}{N}\}. \quad (9)$$

Treating N as a continuous control variable and using calculus, we see that the optimal number of containers N^* that minimizes the port manager's *LRENC* is given by[41]

$$N^*=\sqrt{\frac{2\alpha C\left(1-\alpha E\left[I\right]\right)}{c}}. \quad (10)$$

In words, equation (10) tells us that the optimal number of containers N^* equals the square root of the ratio of the product of twice the Poisson arrival rate of containers (α), the second source of net cost (C), and the term $\left(1-\alpha E\left[I\right]\right)$ to the first source of net cost (c). Looking over equation (10), it is easy to verify two properties of the optimal number of containers N^*. First, as the second source of net cost (C) increases, it is in the interest of the port manager to *raise* the optimal number of containers in the inspection cycle. Second, if the first source of net cost (c) goes up, then the port manager finds it desirable to *lower* the optimal number of containers in the inspection cycle.

41 The reader can check to see that the relevant second order condition is satisfied.

Let us now substitute the expression for the optimal number of containers from equation (10) into the minimand in equation (9). This gives us an expression for the minimal *LRENC* that our port manager will incur by selecting the number of containers in the inspection cycle optimally. Let this minimal *LRENC* be $(LRENC)^{*}_{CP}$, where the subscript *CP* denotes the container policy. Some algebra tells us that

$$(LRENC)^{*}_{CP} = \alpha c E[I] + \frac{\alpha^2 c E\left[I^2\right]}{2\left(1-\alpha E[I]\right)} + \sqrt{2\alpha c C\left(1-\alpha E[I]\right)} - \frac{c}{2} \quad (11)$$

Inspecting equations (8) and (11), we see that the first two terms on the RHSs of these two equations are independent of the optimal number of containers N^* and hence these two terms are identical. However, N^* affects the last two terms on the RHS of equation (8). Hence, when we substitute the optimized value of N, N^*, into equation (8), we get the last two terms on the RHS of equation (11). These last two terms on the RHS of equation (11) depend on the Poisson arrival rate of the containers (α), the first source of net cost (c), the second source of net cost (C), and the expectation of the inspection time random variable I.

We now analyze the case in which the focus of our port manager is not on the optimal number of containers in an inspection cycle, but on the temporal frequency of inspections. After computing the optimal temporal frequency of inspections, we shall compare equation (11) with the corresponding equation for this latter case in which the port manager's focus is on the temporal dimension of the inspection function.

4. The Temporal Policy

Instead of selecting the number of containers optimally, our port manager now pursues an alternate strategy. In particular, whenever there are no containers in the inspection facility, the inspector leaves this facility and he returns only after a fixed time period T has gone by. The specific task at hand now is to choose T optimally to minimize the *LRENC* from inspection activities.

Let us now calculate the port manager's *LRENC* when his focus is on the control variable T, and not on the optimal number of containers. As in the previous section, we suppose that containers arrive at the inspection facility in the port under study in accordance with a Poisson process with rate α. Further, also as in section 3, we shall use the theory of renewal-reward processes in general and the renewal-reward theorem in particular to compute the *LRENC* from inspections.

To this end, let us say that a new cycle begins every time the inspector *leaves* the inspection facility.[42] Also, from the discussion in the previous section, recall that every cycle has a period during which the inspector is busy and a period during which the inspector is idle. Given these two points, we now follow Batabyal *et al.* (2001) and condition on *N(T)*, the number of container arrivals in the time period in which the inspector is away from the inspection facility, to compute the two expectations of interest, i.e., *E*[*net cost per cycle*] and *E*[*length of cycle*]. Conditioning on *N(T)*, we get

$$E[\textit{net cost per cycle}/N(T)] = \sum_{j=1}^{j=N(T)} E\left[\widehat{C}_j\right] + C + \frac{cN(T)T}{2}. \tag{12}$$

We shall now use the following four pieces of information to simplify equation (12) further. First, from equation (6) we get an expression for $E[\widehat{C}_j]$. Second, from theorem 2.3.1 in Ross (1996, p. 67), we can tell that the times at which the containers arrive at the inspection facility are independently and uniformly distributed random variables on the interval (0,*T*). Third, from Ross (1996, pp. 59-60) we conclude that *N(T)* is distributed as a Poisson random variable with mean αT. Finally,

42The reader should note the difference in the meanings of the word "cycle" in this section and the previous section 3. In section 3, a new cycle began every time the inspector returned to the inspection facility. In contrast, in this section, a new cycle begins every time the inspector leaves the inspection facility.

from Ross (2003, p. 519) we reason that $E[N(T)\{N(T)-1\}]=E[N^2(T)]$ - $E[N(T)]=(\alpha T)^2$. Using these four pieces of information to simplify equation (12), we get

$$E[net\ cost\ per\ cycle]=\frac{(\alpha T)^2 cE[I]}{2(1-\alpha E[I])}+\alpha cTE[B_S]+C+\frac{\alpha cT^2}{2}. \quad (13)$$

Our next task is to determine *E*[*length of cycle*]. Once again, conditioning on *N*(*T*) and then using the properties of the expectation operator (see Ross (1996, p. 21)) we get

$$E[length\ of\ cycle]=E[E[length\ of\ cycle/N(T)]]=E[T+N(T)E[B]]. \quad (14)$$

We now focus on the expectation on the extreme RHS of equation (14). Specifically, let us simplify this expectation using equation (3) and the fact that $E[N(T)]=\alpha T$. This gives us

$$E[length\ of\ cycle]=T+\frac{\alpha E[I]T}{1-\alpha E[I]}=\frac{T}{1-\alpha E[I]}. \quad (15)$$

We now divide the RHS of equation (13) by the RHS of equation (15) and then use the result $E[B_S]=\alpha E[I^2]/\{2(1-\alpha E[I])^2\}+E[I]/(1-\alpha E[I])$ to eliminate $E[B_S]$. After some algebra, we get

$$LRENC = \alpha cE[I] + \frac{\alpha^2 cE\left[I^2\right]}{2\left(1-\alpha E\left[I\right]\right)} + \frac{\alpha cT}{2} + \frac{C\left(1-\alpha E\left[I\right]\right)}{T}. \quad (16)$$

Having computed the expression for the *LRENC* of inspections, we are now in a position to state our port manager's optimization problem. Specifically, this manager chooses the time during which the inspector is absent from the inspection facility T to minimize the *LRENC* from inspection activities. Formally, our port manager solves

$$min_{\{T\}}\{\alpha cE[I] + \frac{\alpha^2 cE\left[I^2\right]}{2\left(1-\alpha E\left[I\right]\right)} + \frac{\alpha cT}{2} + \frac{C\left(1-\alpha E\left[I\right]\right)}{T}\}. \quad (17)$$

Using calculus, the optimal time T^* during which the inspector ought to be absent from the inspection facility is given by[43]

$$T^* = \sqrt{\frac{2C\left(1-\alpha E\left[I\right]\right)}{\alpha c}}. \quad (18)$$

In words, T^* equals the square root of the ratio of the product of twice the second source of net cost (C) and the term ($1-\alpha E\left[I\right]$) to the product of the Poisson arrival rate of the containers (α) and the first

43 The reader can check to see that the pertinent second order condition is satisfied.

source of net cost (c). Going over equation (18), it is straightforward to verify two properties of the optimal leave period T^*. First, as the second source of net cost (C) goes up, the port manager's optimal response calls for *lengthening* the time period during which the inspector is away from the inspection facility. In contrast, when the first source of net cost (c) increases, it is optimal to *shorten* the time during which the inspector is absent from the inspection facility.

Let us now substitute the expression for T^* from equation (18) into the minimand in equation (17). This gives us an expression for the optimized *LRENC* from inspections given that the inspector's leave period has been selected optimally. Denote this optimized *LRENC* by $(LRENC)_{TP}^*$ where *TP* denotes the temporal policy. After several steps, we get

$$(LRENC)_{TP}^* = \alpha c E[I] + \frac{\alpha^2 c E\left[I^2\right]}{2\left(1-\alpha E[I]\right)} + \sqrt{2\alpha c C\left(1-\alpha E[I]\right)}. \tag{19}$$

Inspecting equations (16) and (19), we see that the first two terms on the RHSs of these two equations are independent of the optimal leave period T^* and hence these two terms are identical. Even so, T^* impacts the last two terms on the RHS of equation (16). Hence, when we substitute the optimized value of T, T^*, into equation (16), we get the last term on the RHS of equation (19). This last term on the RHS of equation (19) is a function of the Poisson arrival rate of the containers (α) the first source of net cost (*c*), the second source of net cost (*C*), and the expectation of the inspection time random variable *I*.

Recall that the objective of our chapter is to provide an answer to the following question: Is the *LRENC* of inspections lower with the container policy or with the temporal policy? We now provide a precise answer to this question.

5. The Optimal Inspection Policy in Invasive Species Management

Equations (11) and (19) provide us with expressions for the optimized *LRENC* when the number of containers and the leave period have been chosen optimally. Comparing these two expressions, we see that

$$(LRENC)^{*}_{CP} = \alpha cE[I] + \frac{\alpha^2 cE\left[I^2\right]}{2\left(1-\alpha E[I]\right)} + \sqrt{2\alpha cC\left(1-\alpha E[I]\right)} - \frac{c}{2} <$$

$$\alpha cE[I] + \frac{\alpha^2 cE\left[I^2\right]}{2\left(1-\alpha E[I]\right)} + \sqrt{2\alpha cC\left(1-\alpha E[I]\right)} = (LRENC)^{*}_{TP} \qquad (20)$$

Equation (20) clearly tells us that the port manager's *LRENC* with an optimally chosen number of containers in the inspection cycle is *lower* than his *LRENC* with an optimally chosen leave period in the inspection cycle. It is in this sense that the container policy is superior to the temporal policy. Put differently, if the port manager had to choose a single control variable from a control set consisting of the number of containers and the leave period, then this manager would choose the number of containers over the leave period.

The reader will note that the result described in the previous paragraph is based on explicit optimization by the port manager. However, it may not always be possible to conduct such involved optimization exercises. In such situations, the port manager may want to work with simple or "rule of thumb" decision making principles. Therefore, it is of some interest to determine whether the superiority of the container policy result holds when, instead of optimizing, our port manager chooses the leave period for the inspector on the basis of a rule of thumb. In this inspection context, what might a (relatively simple) rule of thumb temporal policy look like? To answer this question, let us reconsider the expressions for the two *LRENC's* in equations (8) and (16).

Inspecting the RHSs of these two equations, we see that the first two terms in both equations are identical. So, no rule of thumb emerges by "eyeballing" these two terms. However, if our port manager were to equate the two third terms in these two equations, i.e., set $c(N-1)/2=\alpha cT/2$, then he would select $T_1=(N-1)/\alpha$ and this is our first rule of thumb. Obviously, this is not the only possibility. If the manager equated the two fourth terms on the RHSs of equations (8) and (16), i.e., set $\alpha C(1-\alpha E[I])/N=C(1-\alpha E[I])/T$, then he would choose $T_2=N/\alpha$ and this is our second rule of thumb. By substituting these values of T_1 and T_2 into equation (16) and then comparing the emerging two equations with equation (8), the reader can verify that our previous result $(LRENC)^*_{CP} < (LRENC)^*_{TP}$ holds for both the above mentioned rules of thumb. From this, we conclude that our basic result about the superiority of the container policy is *robust*. It holds not only when the port manager chooses the inspection policy on the basis of an explicit optimization exercise but also when this manager chooses the inspector's leave period with rules of thumb.

6. Conclusions

Maritime trade today routinely involves the transport of goods by means of containers on ships from one region of the world to another. This transport of goods by means of containers on ships often results in detrimental invasions of new regions by alien plant and animal species. Therefore, if an apposite authority such as a port manager's aim is to prevent biological invasions, then this manager must inspect arriving containers for potentially deleterious biological organisms. Given this state of affairs, what kind of inspection policy ought a manager to optimally have in place? In particular, is the *LRENC* of inspections lower when a port manager follows a policy of inspecting cargo upon the arrival of a specified number of containers (the container policy) or is this *LRENC* lower when this manager inspects cargo at fixed points in time (the temporal policy)? Our analysis shows that *irrespective* of whether the inspection policy choice is made on the basis of an explicit optimization exercise or on the basis of rules of thumb, the container policy is superior to the temporal policy because the container policy results in lower *LRENC* from inspection activities.

The analysis in this chapter can be extended in a number of directions. We now make two suggestions for extending the research described in this chapter. First, given the work of Roberts and Spence (1976), environmental economists and regional scientists now know that a pure price control instrument (fee or tax) and a pure quantity control (emissions permit scheme) instrument can be combined to create a hybrid control instrument that is part-price and part-quantity in nature. Roberts and Spence (1976) showed that this hybrid control instrument can always be converted into a pure price or a pure quantity control instrument. Therefore, in comparison with either a pure price or a pure quantity control instrument, a regulator will typically do at least as well and possibly much better with this hybrid control instrument. Building on this basic Roberts and Spence (1976) insight, it would be useful to ascertain whether it is possible to design a hybrid inspection policy that is part-container and part-temporal in nature. If this can be done, then the next step would be to determine whether this part-container and part-temporal hybrid inspection policy dominates the pure container and the pure temporal inspection policies in terms of, for instance, the *LRENC* criterion.

Second, it would also be interesting to determine whether the results we have obtained in this chapter with the *M/G/*1 queuing model carry over when one analyzes the basic inspection policy choice question of this chapter with alternate queuing models such as the *G/M/*1 model or the *G/G/*1 model. Studies of invasive species management that incorporate these aspects of the inspection policy choice question into the analysis will provide additional insights into a management problem that has considerable economic and biological consequences.

References

Barbier, E.B. 2001. A note on the economics of biological invasions, *Ecological Economics*, 39, 197-202.

Batabyal, A.A., Biswas, B., and Godfrey, E.B. 2001. On the choice between the stocking rate and time in range management, *Environmental and Resource Economics*, 20, 211-223.

Batabyal, A.A. 2004. A research agenda for the study of the regulation of invasive species introduced unintentionally via maritime trade, *Journal of Economic Research*, 9, 191-216.

Batabyal, A.A., and Beladi, H. 2006. International trade and biological invasions: A queuing theoretic analysis of the prevention problem, *European Journal of Operational Research*, 170, 758-770.

Batabyal, A.A., Beladi, H., and Koo, W.W. 2005. Maritime trade, biological invasions, and the properties of alternate inspection regimes, *Stochastic Environmental Research and Risk Assessment*, 19, 184-190.

Cox, G.W. 1993. *Conservation Ecology*. W.C. Brown Publishers, Dubuque, IA.

Eiswerth, M.E., and Johnson, W.S. 2002. Managing nonindigenous invasive species: Insights from dynamic analysis, *Environmental and Resource Economics*, 23, 319-342.

Eiswerth, M.E., and van Kooten, G.C. 2002. Uncertainty, economics, and the spread of an invasive plant species, *American Journal of Agricultural Economics*, 84, 1317-1322.

Heywood, V. Ed. 1995. *Global Biodiversity Assessment*. Cambridge University Press, Cambridge, UK.

Horan, R.D., Perrings, C., Lupi, F., and Bulte, E.H. 2002. Biological pollution prevention strategies under ignorance: The case of invasive species, *American Journal of Agricultural Economics*, 84, 1303-1310.

Kulkarni, V.G. 1995. *Modeling and Analysis of Stochastic Systems*. Chapman and Hall, London, UK.

Office of Technology Assessment (OTA). 1993. *Harmful Non-Indigenous Species in the United States*. OTA-F-565, Washington, DC.

Olson, L.J., and Roy, S. 2002. The economics of controlling a stochastic biological invasion, *American Journal of Agricultural Economics*, 84, 1311-1316.

Parker, I.M., Simberloff, D., Lonsdale, W.M., Goodell, K., Wonham, M., Kareiva, P.M., Williamson, M.H., Von Holle, B., Moyle, P.B., Byers, J.E., and Goldwasser, L. 1999. Impact: Toward a framework for understanding the ecological effects of invaders, *Biological Invasions*, 1, 3-19.

Perrings, C., Williamson, M., and Dalmazzone, S. 2000. Introduction in C. Perrings, M. Williamson, and S. Dalmazzone, (Eds.), *The Economics of Biological Invasions*. Edward Elgar, Cheltenham, UK.

Pimentel, D., Lach, L., Zuniga, R., and Morrison, D. 2000. Environmental and economic costs of nonindigenous species in the United States, *BioScience*, 50, 53-65.

Roberts, M.J., and Spence, A.M. 1976. Effluent charges and licenses under uncertainty, *Journal of Public Economics*, 5, 193-208.

Ross, S.M. 1996. *Stochastic Processes*, 2nd edition. Wiley, New York, NY.

Ross, S.M. 2002. *Probability Models for Computer Science*. Harcourt Academic Press, San Diego, CA.

Ross, S.M. 2003. *Introduction to Probability Models*, 8th edition. Academic Press, San Diego, CA.

Taylor, H.M., and Karlin, S. 1998. *An Introduction to Stochastic Modeling*, 3rd edition. Academic Press, San Diego, CA.

USDA, APHIS, and USFS. 2000. Pest Risk Assessment for Importation of Solid Wood Packing Materials into the United States. http://www.aphis.usda.gov/ppq/pra/swpm/complete.pdf

Vitousek, P.M., D'Antonio, C.M., Loope, L.L., and Westbrooks, R. 1996. Biological invasions as global environmental change, *American Scientist*, 84, 468-478.

Wolff, R.W. 1989. *Stochastic Modeling and the Theory of Queues*. Prentice-Hall, Englewood Cliffs, NJ.

Yang, Z., and Perakis, A.N. 2004. Multiattribute decision analysis of mandatory ballast water treatment measures in the US great lakes, *Transportation Research Part D*, 9, 81-86.

CHAPTER 8

SOME STATISTICAL PROPERTIES OF A GENERIC CONTAINER INSPECTION POLICY IN INVASIVE SPECIES MANAGEMENT - WITH S.J. YOO

Abstract. In recent times—see Batabyal and Nijkamp (2005) and Batabyal and Lee (2006)—there has been considerable interest in studying container inspection policies as a regulatory tool in invasive species management. In this chapter, we continue this line of inquiry and first delineate a generic container inspection policy. Next, we derive the generating function for the random number of containers in a seaport when this generic policy is used to inspect containers. Finally, we first compute the mean and the variance of the limiting distribution of the number of containers in the seaport under study and then we discuss three ways in which these two moments might be used to manage invasive species effectively.

1. Introduction

In this era of globalization, airplanes, trucks, and particularly ships, are commonly used to transport goods from one region of the world to another. The work of Parker *et al.* (1999), Batabyal (2004), and Work *et al.* (2005) tells us that in addition to transporting goods between regions, airplanes, trucks, and ships have unwittingly also managed to transport a whole host of invasive plant and animal species (also referred to as alien or non-native species) from one geographical region of the world to another.

This unwitting transportation has occurred in many different ways. Here are three examples. First, invasive animal species have occasionally succeeded in lodging themselves in the landing gear of airplanes and, in this way, they have traveled as stowaways from one region to another. Second, a number of marine invasive species have been introduced unintentionally into a region by ships dumping their ballast water. Cargo ships often carry ballast water in order to increase vessel stability when they are not carrying full loads. When these ships come into a seaport, this ballast water must be discarded before cargo can be loaded. Finally, ships and trucks have introduced invasive species into a particular region by means of the containers that they regularly use to transport cargo from one region to another. It is important to comprehend that invasive species can remain hidden in containers for extended periods of time. In addition, the material—such as wood—that is often used to pack the cargo in the containers may itself contain invasive species.

Invasive species have often succeeded in invading new habitats and the ensuing biological invasions have turned out to be very costly for society. For the United States alone, the magnitude of these costs is enormous. For instance, the Office of Technology Assessment (OTA

(1993)) has estimated that the Russian wheat aphid caused $600 million worth of crop damage between 1987 and 1989. In addition to these economic costs, invasive species have given rise to serious biological damage as well. In this regard, Vitousek *et al.* (1996) have demonstrated that invasive species can alter ecosystem processes, act as vectors of diseases, and diminish biological diversity. Further, Cox (1993) has remarked that out of 256 vertebrate extinctions with a known cause, 109 are the result of biological invasions. The upshot of this discussion is clear: Invasive species have often posed a great risk to nations.

Generally speaking, actions intended to control the spread of invasive species and their deleterious effects are either *pre-invasion* or *post-invasion* in nature. The goal of pre-invasion actions is to preclude invasive species from overrunning a new region. In contrast, post-invasion actions involve the optimal control of an invasive species, given that this species has already invaded a new region.

Recently, there has been great interest in studying one kind of pre-invasion action, namely, container inspection policies. For instance, Batabyal and Nijkamp (2005) have shown that, in an inspection cycle, irrespective of whether the inspection policy choice is made on the basis of an optimization exercise or on the basis of a rule of thumb, the "container policy" dominates the "temporal policy" because the container policy results in lower long run expected net cost (*LRENC*) from inspections. The container policy involves the inspection of cargo upon the arrival of a specified number of containers in a seaport. In contrast, the temporal policy involves the inspection of container cargo at fixed points in time. Similarly, Batabyal and Lee (2006) have analyzed container inspection policies in three scenarios. In the first scenario, the time taken to conduct inspections is negligible. In the second scenario, this inspection time is deterministic. In the third and final scenario, this same inspection time is stochastic. The authors compare and contrast the *LRENC* and the optimal container inspection policy in each of these three scenarios.

In this chapter, we continue the study of container inspection policies and the rest of this chapter is organized as follows. In section 2.1 we delineate a generic container inspection policy. In section 2.2, we derive the generating function for the random number of containers in a seaport when the above generic policy is used to inspect containers in a seaport. In section 2.3, we first compute the mean (m) and the variance (v) of the limiting distribution of the number of containers in the seaport, and then we discuss three ways in which these two moments might be used to manage invasive species effectively. Section 3 concludes and offers suggestions for future research on the subject of this chapter.

2. The Theoretical Model

2.1. A generic container inspection policy

Suppose that batches of containers from individual ships arrive in a seaport in accordance with a Poisson process with rate $\alpha>0$. The number of containers in a batch or the batch size is an independent random variable with a common discrete probability distribution $\{\delta_j, j=1,2,3,\ldots\}$ with finite second moment. Our seaport is large, in the sense that it either actually has or can make available space to store any number of containers. Therefore, the arriving containers are physically stored in this seaport.

The container inspection policy of interest in this chapter is generic in the sense that the inspection times of the various containers in the seaport are independent random variables with a *general* probability distribution $B(x)$ with finite first moment μ. In addition, the inspection times of containers from the same batch, i.e., from the same ship, are independent of each other. Let $\delta(z)=\sum_{j=1}^{\infty}\delta_j z^j$ be the generating function[44] of the batch size and let $\{p_j\}$ be the limiting distribution of the number of containers present in our seaport.

As described in Batabyal *et al.* (2005), long delays in inspecting and clearing containers before their contents can be transported to an

44For more on generating functions and their properties, see Taylor and Karlin (1998, pp. 184-187) or Tijms (2003, pp. 449-455).

interior point of the region in which our seaport is located gives rise to economic costs such as the costs that arise from the temporary stoppage of economic activities in the seaport under study. Therefore, when the previous paragraph's generic policy is used to inspect arriving containers for potentially deleterious invasive species, one question that assumes considerable significance concerns the *number* of containers present in our seaport.[45] This number is obviously a non-negative and integer valued random variable and hence its essential statistical properties can be obtained by deriving the generating function associated with this number. Therefore, we now proceed to this derivation.

2.2. Derivation of the generating function

Let $P(z)$ denote the pertinent generating function. The reader should note that it makes sense to focus on this generating function because "a discrete probability distribution is uniquely determined by its generating function" (Tijms, 2003, p. 449). Now, from equation 9.2 in Taylor and Karlin (1998, p. 185) it follows that

$$P(z)=\sum_{j=0}^{\infty} p_j z^j . \quad (1)$$

45In this regard, the reader should note that the economic costs that we have just alluded to are an increasing function of the number of containers present in our seaport.

Given the information in section 2.1, the task before us now is to simplify the right-hand-side (RHS) of equation (1).

To this end, suppose that our seaport is empty at time 0. Let $X(t)$ denote the number of inspectors who are busy doing their jobs at time t, and let $P(t, z)=E[z^{X(t)}]$ where $E[\bullet]$ is the expectation operator. Now, by conditioning on what might transpire in the small time interval $(0, \Delta t)$ we can tell that

$$P(t+\Delta t,z)=E[z^{X(t+\Delta t)}]=(1-\alpha\Delta t)E[z^{X(t)}]+(1-\alpha\Delta t)E[z^{X(t)+C t)}]+o(\Delta t). \tag{2}$$

We now note two things about equation (2). First, $C(t)$ is the number of containers that belong to the batch arriving in the small time interval $(0, \Delta t)$ and that are still being inspected at a time t later. Second, as noted in Tijms (2003, p. 6), the expression $o(\Delta t)$ refers to an unspecified term that is negligibly small compared to Δt itself as $\Delta t \rightarrow 0$. Now, by conditioning on the number of containers or the size of this batch, we can tell that $C(t, z)=E[z^{C(t)}]$ is given by

$$\sum_{l=1}^{\infty}\delta_l\sum_{k=0}^{l}\left(\frac{l}{k}\right)[1-B(l)]^k\,[B(t)]^{l-k}\,z^k=$$

$$\sum_{l=1}^{\infty} \delta_l [B(t)+(1\text{-}B(t))z)]^l = \delta \{B(t)+(1\text{-}B(t))z\}. \qquad (3)$$

We know that the random variables $X(t)$ and $C(t)$ are independent. This allows us to conclude that

$$P(t+\Delta t, z)=(1-\alpha\Delta t)P(t, z)+(1-\alpha\Delta t)P(t, z)C(t, z)+o(\Delta t). \qquad (4)$$

Differentiating and using equation (4), we deduce that $P^{/}(t;z)=-\alpha(1-C(t,z))P(\text{t}, \text{z})$ and, in turn, this gives us

$$P(t;z)=\exp[-\alpha\int_{0}^{t}[1-\delta\{B(x)+1\text{-}B(x)z\}]dx], t>0. \qquad (5)$$

Now, if we let $t\rightarrow\infty$ then we get the generating function we seek and that function is

$$P(z)=\exp[-\alpha\int_{0}^{\infty}\{1-\delta((1\text{-z})\text{B}(x)+z)\}dx]. \qquad (6)$$

Inspection of equation (6) tells us that, *inter alia*, this generating function depends on the Poisson arrival rate of the batches of containers (α), the discrete probability distribution of the batch sizes (δ), and the general probability distribution function of the inspection times of the various containers in the seaport $B(x)$. This concludes the derivation of the generating function of the (random) number of containers in our seaport. We now compute the mean (m) and the variance (v) of the limiting distribution of the number of containers in our seaport.

2.3. Mean and variance

Having derived the key generating function in equation (6), we can easily obtain the first two moments that we seek by appropriately differentiating the above generating function. To this end, let X denote the random container batch sizes. It is clear that X has the discrete probability distribution $\{\delta_j, j=1,2,3\ldots\}$ that we specified in section 2.1. Now, differentiating equation (6) and then evaluating the resulting expression at $z=1$ gives

$$m = P'(1) = \alpha E[X]\mu. \tag{7}$$

Similarly, by differentiating the generating function a second time, we get an expression for the variance and that expression is

$$v = P''(1) + P'(1) - \{P'(1)\}^2 = \alpha E[X]\mu + \alpha E[X(X-1)] \int_0^\infty \{1 - B(x)\}^2 dx. \tag{8}$$

The mean (m) and the variance (v) that we have just computed can be used in invasive species management in three possible ways. First, the mean number of containers in the seaport can be used as an objective function by itself. When this is done, one course of action for a seaport manager might be to minimize the mean number of containers by attempting to alter the probability distribution function $B(x)$ and/or the value of the first moment μ. Second, the mean and the variance of the random number of containers can be part of a more general objective function that a seaport manager might want to optimize.

Finally, the mean and/or the variance can enter an optimization problem—to be solved by the seaport manager—as a constraint. One idea here would be to conduct inspections in such a way so that the net social benefit from inspections is maximized subject to the constraint that the mean number of containers in the seaport under study is always less than some exogenously given threshold, say, $\bar{m}$. Alternately, if greater predictability in the inspection function is desired—we are now thinking of lower variance (v) as a proxy for increased predictability—then one could formulate and solve a net social benefit from inspections maximization problem subject to, for instance, two constraints. The first constraint would tell us that the number of containers in the port under study is at most $\bar{m}$ (as before) and the second constraint would tell us that the variance of the number of containers is also no more than some external threshold, say, $\bar{v}$.

3. Conclusions

In this chapter, we continued a line of inquiry begun recently by Batabyal and Nijkamp (2005) and by Batabyal and Lee (2006). Specifically, we first described a generic container inspection policy. Next, we derived the generating function of the random number of containers in a seaport, when the above mentioned generic policy is used to inspect containers. Finally, we computed the mean and the variance of the limiting distribution of the number of containers in the seaport and then we discussed three ways in which these two moments might be used in invasive species management.

Given the recent controversy—see Anonymous (2006)—about the management of US ports by a firm based in Dubai, the reader should note that the generating function approach that we have studied in this chapter applies not only in the context of invasive species management, but more generally, as well. In fact, this approach is relevant in any context, such as the control of illegal goods and the interdiction of weapons of mass destruction that involves the inspection of containers brought in on ships.

The analysis in this chapter can be extended in a number of directions. Here are two suggestions for extending the research described here. First, following the discussion towards the end of section 2.3, it would be useful to set up and solve optimization problems involving the efficient allocation of inspection resources by a seaport manager. Second, to account for the fact that batches of containers are more likely to arrive at a seaport at certain times than at other times, it would be helpful to model the arrival of the batches of containers with a non-homogenous Poisson process. Studies of inspections in invasive species management that incorporate these features of the problem into the analysis will provide further insights into a management task that has salient economic and ecological implications.

References

Anonymous. 2006. From Karl Marx's copybook, *The Economist*, March 2.

Batabyal, A.A. 2004. A research agenda for the study of the regulation of invasive species introduced unintentionally via maritime trade, *Journal of Economic Research*, 9, 191-216.

Batabyal, A.A., Beladi, H., and Koo, W.W. 2005. Maritime trade, biological invasions, and the properties of alternate inspection regimes, *Stochastic Environmental Research and Risk Assessment*, 19, 184-190.

Batabyal, A.A., and Nijkamp, P. 2005. On container versus time based inspection policies in invasive species management, *Stochastic Environmental Research and Risk Assessment*, 19, 340-347.

Batabyal, A.A., and Lee, D.M. 2006. The infinitesimal, the deterministic, and the probabilistic: Alternate container inspection policies in invasive species management, *Biological Invasions*, 8, 1663-1671.

Cox, G.W. 1993. *Conservation Ecology*. W.C. Brown Publishers, Dubuque, IA.

Office of Technology Assessment (OTA). 1993. *Harmful Non-Indigenous Species in the United States*. OTA-F-565, Washington, DC.

Parker, I.M., Simberloff, D., Lonsdale, W.M., Goodell, K., Wonham, M., Kareiva, P.M., Williamson, M.H., Von Holle, B., Moyle, P.B., Byers, J.E., and Goldwasser, L. 1999. Impact: Toward a framework for understanding the ecological effects of invaders, *Biological Invasions*, 1, 3-19.

Taylor, H.M., and Karlin, S. 1998. *Introduction to Stochastic Modeling*, 3rd edition. Academic Press, San Diego, CA.

Tijms, H.C. 2003. *A First Course in Stochastic Models*. Wiley, Chichester, UK.

Vitousek, P.M., D'Antonio, C.M., Loope, L.L., and Westbrooks, R. 1996. Biological invasions as global environmental change, *American Scientist*, 84, 468-478.

Work, T.T., McCullough, D.G., Cavey, J.F., and Komsa, R. 2005. Arrival rate of nonindigenous insect species into the United States through foreign trade, *Biological Invasions*, 7, 323-332.

CHAPTER 9

SOME PROBABILISTIC ATTRIBUTES OF INSPECTION POLICIES IN ALIEN SPECIES MANAGEMENT - WITH P. NIJKAMP

Abstract. Inspections are a salient part of alien species management and, therefore, in contemporary times, there has been great interest in analyzing the role of inspections as an apposite regulatory tool. In this chapter, we continue the study of inspections in a probabilistic environment and we shed light on three specific issues concerning inspection policies in the context of alien species management. First, given a particular seaport, we determine the likelihood that $n \in \mathbb{N}$ ships will need to be inspected in a relevant time period. Second, we compute the probability distribution function of the amount of time it takes to inspect an arbitrary ship. Finally, we show how the probability distribution function of the total cost of inspections in the above time period can be calculated.

1. Introduction

We live in an era of globalization. The phenomena of globalization in general, and the dreadful events of 11 September 2001 in particular, have given rise to great interest in issues related to security across of the world. In the United States (US), one way in which this interest has manifested itself is in the sharply heightened interest in inspecting goods that are transported into the country from other parts of the world by means of airplanes, trucks, and particularly ships. This concern with seaport security in particular is not misplaced. As *The Economist* (Anonymous, 2006) has recently pointed out, only about five percent of the containers that bring two billion tonnes of cargo to US ports are actually inspected. As a result, it is not difficult at all for all kinds of illegal goods and—what's important for our purpose—potentially deleterious animal and plant species to get into the US.

The research of Parker *et al.* (1999), Batabyal (2004), and Work *et al.* (2005) is unequivocal on the point that, in addition to carrying goods between regions, airplanes, trucks, and ships have unintentionally also managed to carry all kinds of alien plant and animal species (also referred to as invasive or non-native species) from one geographical region of the world to another. This unintentional carriage has taken place in many different ways. Here are three examples: First, every now and then, alien animal species have managed to lodge themselves in the landing gear of airplanes and, in this way, they have traveled as stowaways from one region to another. Second, a number of marine alien species have been introduced unintentionally into a region by ships dumping their ballast water. Cargo ships frequently carry ballast water in order to enhance vessel stability when they are not carrying full loads. When these ships come into a seaport, this ballast water must be jettisoned before cargo can be loaded. Finally, and perhaps most importantly, ships and trucks have introduced alien species into a particular region by means of the containers that they customarily use to carry cargo from one region to another. It is important to understand that alien species can remain hidden in containers for long periods of time. In addition, the material—such as wood—that is frequently used to pack the cargo in the containers may itself contain alien species.

When alien species have succeeded in invading new habitats, the resulting biological invasions have often resulted in great losses for society. For the US alone, the magnitude of these losses is enormous. As an example, the Office of Technology Assessment (OTA (1993)) has calculated that the Russian wheat aphid caused $600 million worth of crop damage between 1987 and 1989. In addition to these economic losses, alien species have also given rise to serious biological damage. In this regard, the research of Vitousek *et al.* (1996) is pertinent. These researchers have shown that alien species can change ecosystem processes, act as vectors of diseases, and reduce biological diversity. Further, the research of Cox (1993) tells us that, out of 256 vertebrate extinctions with a known cause, 109 are the outcome of biological invasions. The main message from the discussion in this paragraph is clear: Alien species have often imposed great costs on the affected nations.

Managerial actions intended to control the spread of alien species and their injurious effects are, broadly speaking, either *pre-invasion* or *post-invasion* in nature. The objective of pre-invasion actions is to prevent alien species from invading a new region. In contrast, post-invasion actions are designed to optimally control an alien species, given that this species has already invaded a new region. Recently, there has been great interest in analyzing one kind of pre-invasion action, i.e., container inspection policies. In this regard, Batabyal and Nijkamp (2005) have shown that, in an inspection cycle, the "container policy" is superior to the "temporal policy," because the container policy results in lower long run expected net cost (*LRENC*) from inspections.[46] Similarly, Batabyal and Lee (2006) have compared and contrasted the *LRENC* and the optimal container inspection policy in three scenarios. In the first scenario, the time taken to conduct inspections is negligible. In the second scenario, this inspection time is deterministic. In the third scenario, this same inspection time is stochastic. Finally, Batabyal and Yoo (2006) have studied the statistical properties of what they call a generic container inspection policy.

Following Batabyal and Yoo (2006), we continue the probabilistic study of inspection policies and the rest of this chapter is

46The container policy involves the inspection of cargo upon the arrival of a particular number of containers in a seaport. In contrast, the temporal policy involves the inspection of container cargo at fixed points in time.

organized as follows. In section 2.1, given a particular seaport, we determine the *likelihood* that $n \in \mathbb{N}$ ships (carrying container cargo) will need to be inspected in a particular time period. In section 2.2, we compute the probability distribution function of the amount of time it takes to inspect an *arbitrary* ship. In section 2.3, we show how the probability distribution function of the *total cost* of inspections in the above time period can be calculated. Section 3 concludes and offers suggestions for future research on the subject of this chapter.

2. The Theoretical Model

2.1. Likelihood of inspecting n ships

Following Batabyal (2006), we analyze a scenario in which there are two exporting countries and one importing country called Home. The subsequent analysis is conducted from the perspective of the manager of a representative seaport in Home. Home imports goods from two separate countries and these two exporting nations transport their goods via ships to the seaport under study in Home. Ships from the two exporting countries carry their container cargo into the seaport in Home. The Home seaport manager's concern is that the container cargo may or may not contain potentially injurious alien animal and/or plant species. Therefore, this seaport manager's job is to ensure, to the extent possible, that all the container cargo arriving in ships from the two exporting nations in his seaport is inspected for the presence of potentially deleterious alien species.

Consistent with the approach adopted in Batabyal *et al.* (2005) and Batabyal and Lee (2006), ships from the two exporting countries A and B arrive at our seaport in Home in accordance with independent Poisson processes with rates λ_A and λ_B per time period.[47] In addition, we suppose that the time it takes to inspect the ships from the two countries A and B is exponentially distributed with means $1/\mu_A$ and $1/\mu_B$. The task before us now is to determine the likelihood that $n \in \mathbb{N}$ ships will need to be inspected in the next time period of interest.[48]

To compute the above likelihood or $Prob\{inspect\ n\ ships\}$, let us first use theorem 1.1.3 in Tijms (2003, p. 6) and recognize that the sum of two independent Poisson processes is also a Poisson process. In addition, it is well known—see Ross (1996, p. 64)—that the Poisson process is memoryless. These two facts together tell us that the likelihood we seek is given by

47 See Ross (1996, pp. 59-97) and Tijms (2003, pp. 1-32) for textbook accounts of the Poisson process.

48 Although we shall not always be explicit about this point, the reader should understand that, when we talk about inspecting a ship, we are really talking about inspecting the containers on that ship.

$$Prob\{inspect\ n\ ships\} = e^{-(\lambda_A+\lambda_B)} \frac{(\lambda_A+\lambda_B)^n}{n!},\ n=0,1,2,... \tag{1}$$

The number of ships that will need to be inspected in our seaport in the next time period of interest turns out to be a Poisson distributed random variable. Therefore, the *likelihood* that $n \in \mathbb{N}$ ships will need to be inspected in this time period is given by equation (1). This completes the first of three tasks that we have set for ourselves in this chapter. We now proceed to compute the probability distribution function of the amount of time it takes to inspect an arbitrary ship.

2.2. Arbitrary ship probability distribution function

To undertake the requisite computation, it will be necessary to first determine the probability that the next ship arriving at the seaport under study is from either country *A* or *B*. Because we are studying a two exporting country model, some thought tells us that the two probabilities we seek are given by $\lambda_A/(\lambda_A+\lambda_B)$ and by $\lambda_B/(\lambda_A+\lambda_B)$. With this information, we can reason that the probability distribution function of the amount of time it takes to inspect an arbitrary ship is given by

$$\frac{\lambda_A}{\lambda_A+\lambda_B} f_A(x) + \frac{\lambda_B}{\lambda_A+\lambda_B} f_B(x), \tag{2}$$

where $f_i(x) = \mu_i e^{-\mu_i x}$, $i=A,B$.

In other words, the probability distribution function we seek is the weighted sum of two distinct exponential density functions where the weights are the probabilities that an arbitrary ship is either from country *A* or from country *B*. Equation (2) provides our seaport manager in Home with one way of allocating scarce inspection resources in his seaport along the temporal dimension. Because these inspection resources are scarce, they are costly to use. This brings us to the third and final task of this chapter, and this task involves showing how we can calculate the probability distribution function of the total cost of inspections in a relevant time period.

2.3. Total cost probability distribution function

As in section 2.1, ships from the two exporting countries *A* and *B* arrive in the seaport under study in Home in accordance with independent Poisson processes with rates λ_A and λ_B per time period. Now, each time a ship from country *A* (*B*) arrives, we suppose that it costs C_A (C_B) dollars to inspect this ship for the possible presence of alien animal and/or plant species in its container cargo. In other words, a maintained assumption in our analysis in this section is that all ships from the same country are equally costly to inspect, but ships from different countries give rise to different inspection costs because $C_A \neq C_B$.

We have already noted in section 2.1 that the sum of two independent Poisson processes is also a Poisson process. Therefore, using definition 1.2.1 in Tijms (2003, p. 18), it follows that the total cost of inspections in t time units is a compound Poisson variable and the rate of the pertinent Poisson process is $\lambda = \lambda_A + \lambda_B$.[49] Further, the distributions of the independent and identically distributed non-negative random variables D_1, $D_2 \ldots$ specified in definition 1.2.1 are given by the following expressions. Specifically, we have

$$a_j = Prob\{D_1 = j\} = \frac{\lambda_A}{\lambda_A + \lambda_B}, j = c_A,\ a_j = \frac{\lambda_B}{\lambda_A + \lambda_B}, j = c_B,\ a_j = 0 \ otherwise \tag{3}$$

49 See Ross (1996, pp. 82-88) and Tijms (2003, pp. 18-22) for more details on compound Poisson random variables and processes.

We can now use theorem 1.2.1 in Tijms (2003, p. 19) and conclude that the generating function[50] of the total cost of inspections in t time units is given by

$$R(z,t)= e^{-\lambda t\{1-A(z)\}},\ |z|\leq 1, \tag{4}$$

where $A(z)=\sum_{j=0}^{\infty} a_j z^j$, $|z|\leq 1$, and the various $a_j^{\prime}s$ in the previous expression for $A(z)$ are given in equation (3). Tijms (2003, p. 19 and pp. 449-450) has noted that, if we know the generating function of a particular random variable then, for all practical purposes, we also know its probability distribution function. Therefore, we have now shown how to calculate the probability distribution function of the total cost of container inspections in the relevant time period.

The information contained in equations (3) and (4) can be used to facilitate the general task of alien species management in three possible ways. First, equations (3) and (4) can be used to compute the expected total cost of inspections in the relevant time period of interest. When this is done, one course of action for our seaport manager might be to choose an apposite control variable to minimize this expected total cost. Second, this expected total cost can also be part of a more general objective

50See Tijms (2003, pp. 449-455) for additional details on the generating function.

function—such as an expected net social benefit from inspections function—that a seaport manager might want to maximize. Finally, the likelihood that $n \in \mathbb{N}$ ships will need to be inspected in the relevant time period (see equation (1)) can play the role of a constraint in either a constrained expected total cost minimization problem or a constrained expected net social benefit maximization problem. Here, the idea would be to conduct container inspections efficiently and in such a way so that, probabilistically speaking, $n \in \mathbb{N}$ ships are inspected in the relevant time period of interest.

3. Conclusions

In this chapter, we continued a line of inquiry begun recently by Batabyal and Nijkamp (2005), by Batabyal and Lee (2006), and by Batabyal and Yoo (2006). Specifically, for the seaport under study, we determined the *likelihood* that $n \in \mathbb{N}$ ships will need to be inspected in a particular time period. Next, we computed the probability distribution function of the amount of time it takes to inspect an *arbitrary* ship from one or the other exporting country. Finally, we showed how the probability distribution function of the *total cost* of inspections in the time period of interest can be calculated and then we discussed three ways in which the findings of this chapter can be used to facilitate the general task of alien species management.

The analysis in this chapter can be extended in a number of directions. Here are two suggestions for extending the research described in this chapter. First, it would be useful to generalize the analysis in this chapter to the case in which there are not two but $m>2$ exporting nations. Second, following the discussion towards the end of section 2.3, it would be useful to set up and solve optimization problems involving the efficient allocation of inspection resources by a seaport manager. Studies of inspection policies in alien species management that incorporate these features of the problem into the analysis will provide further insights into a management task that has salient economic and ecological ramifications.

References

Anonymous. 2006. Trouble on the waterfront, *The Economist*, February 23.

Batabyal, A.A. 2004. A research agenda for the study of the regulation of invasive species introduced unintentionally via maritime trade, *Journal of Economic Research*, 9, 191-216.

Batabyal, A.A. 2006. A rationale for the differential regulatory treatment of imports when invasive species are a potential problem, *Studies in Regional Science*, 36, 179-187.

Batabyal, A.A., Beladi, H., and Koo, W.W. 2005. Maritime trade, biological invasions, and the properties of alternate inspection regimes, *Stochastic Environmental Research and Risk Assessment*, 19, 184-190.

Batabyal, A.A., and Nijkamp, P. 2005. On container versus time based inspection policies in invasive species management, *Stochastic Environmental Research and Risk Assessment*, 19, 340-347.

Batabyal, A.A., and Lee, D.M. 2006. The infinitesimal, the deterministic, and the probabilistic: Alternate container inspection policies in invasive species management, *Biological Invasions*, 8, 1663-1671.

Batabyal, A.A., and Yoo, S.J. 2006. Some statistical properties of a generic container inspection policy in invasive species management, *Ecological Economics*, 60,1-4.

Cox, G.W. 1993. *Conservation Ecology*. W.C. Brown Publishers, Dubuque, IA.

Office of Technology Assessment (OTA). 1993. *Harmful Non-Indigenous Species in the United States*. OTA-F-565, Washington, DC.

Parker, I.M., Simberloff, D., Lonsdale, W.M., Goodell, K., Wonham, M., Kareiva, P.M., Williamson, M.H., Von Holle, B., Moyle, P.B., Byers, J.E., and Goldwasser, L. 1999. Impact: Toward a framework for understanding the ecological effects of invaders, *Biological Invasions*, 1, 3-19.

Ross, S.M. 1996. *Stochastic Processes*, 2nd edition. Wiley, New York, NY.

Tijms, H.C. 2003. *A First Course in Stochastic Models*. Wiley, Chichester, UK.

Vitousek, P.M., D'Antonio, C.M., Loope, L.L., and Westbrooks, R. 1996. Biological invasions as global environmental change, *American Scientist*, 84, 468-478.

Work, T.T., McCullough, D.G., Cavey, J.F., and Komsa, R. 2005. Arrival rate of nonindigenous insect species into the United States through foreign trade, *Biological Invasions*, 7, 323-332.

CHAPTER 10

THE INFINITESIMAL, THE DETERMINISTIC, AND THE PROBABILISTIC: ALTERNATE CONTAINER INSPECTION POLICIES IN INVASIVE SPECIES MANAGEMENT - WITH D.M. LEE

Abstract. Recently, Batabyal and Nijkamp (2005) have shown that, in an inspection cycle, regardless of whether the inspection policy choice is made on the basis of an optimization exercise, or on the basis of a rule of thumb, the "container policy" dominates the "temporal policy" because the container policy results in lower long run expected net cost (*LRENC*) from inspections. In this chapter, we continue this line of inquiry and analyze container policies in three scenarios. In the first scenario the time taken to conduct inspections is negligible. In the second scenario, this inspection time is deterministic and, in the third scenario, this inspection time is stochastic. Specifically, we compare and contrast the *LRENC* and the optimal container policy in each of these three scenarios.

1. Introduction

In contemporary times, airplanes, ships, and trucks have all been used to transport goods from one part of the world to another. Standard textbooks in international economics, such as Feenstra (2004) and

Krugman and Obstfeld (2005), tell us that unfettered trade between different regions of the world is welfare enhancing. Even so, analysts have increasingly pointed out that the magnitude of these welfare gains is likely to be less than what most researchers have hitherto postulated. Why? As Parker *et al.* (1999), Batabyal (2004), and Work *et al.* (2005) have pointed out, this is because in addition to transporting goods between regions, airplanes, ships, and trucks have also managed to transport a whole host of non-native plant and animal species (also known as alien or invasive species) from one geographical region of the world to another.

There are many ways in which airplanes, ships, and trucks have transported non-native plant and animal species from one part of the world to another. Non-native animal species have sometimes managed to lodge themselves in the landing gear of airplanes and, as such, they have traveled as stowaways from one country to another. Similarly, a number of marine non-native species have been introduced unintentionally into a region by ships dumping their ballast water. Cargo ships frequently carry ballast water, in order to increase vessel stability when they are not carrying full loads. When these ships come into port, this ballast water must be released before cargo can be loaded. This means of species introductions is salient and, very recently, the problem of managing alien species that have been introduced into a particular region by means of the dumping of ballast water has received some attention in the literature.[51]

Ships and trucks have introduced non-native species into a particular region by means of the containers that they frequently use to transport cargo from one region to another. In this regard, it is important to understand that alien species can remain undetected in containers for extended periods of time. In addition, the material (such as wood) that is used to pack the cargo in the containers may itself contain invasive species. In the case of maritime shipments, a joint report from the United States Department of Agriculture (USDA), the Animal and Plant Health Inspection Service (APHIS), and the United States Forest Service (USFS) has observed that roughly 51.8% of such shipments contain solid wood packing materials and that infection rates for solid wood packing materials are significant (USDA, APHIS, and USFS (2000, p. 25)). As an illustration, inspections of wooden spools from China uncovered infection rates between 22% and 24% and inspections of braces for granite blocks imported into Canada were found to contain live insects 32% of the time (USDA, APHIS, and USFS (2000, pp. 27-28)).

51 See Yang and Perakis (2004), Batabyal *et al*. (2005), and Batabyal and Beladi (2006) for more on this literature.

The alien species that we have been discussing generally thus far have often managed to invade new habitats and the resulting biological invasions have turned out to be very costly for society. For the United States alone, the dollar value of these costs is stupendous. In this regard, the Office of Technology Assessment (OTA (1993)) has estimated that the Russian wheat aphid caused \$600 million worth of crop damage between 1987 and 1989. Similarly, Pimentel *et al.* (2000) have calculated the total costs of all non-native species at around \$137 billion per year. In addition to these economic costs, non-native species have given rise to serious biological damage as well. For instance, Vitousek *et al.* (1996) have shown that alien species can change ecosystem processes, act as vectors of diseases, and decrease biological diversity. Further, Cox (1993) has pointed out that out of 256 vertebrate extinctions with a known cause, 109 are the outcome of biological invasions. The ramification of the discussion in this paragraph is clear: Invasive species have often been a great menace to society.

A manager interested in addressing the problem of biological invasions can take actions that are either *pre-invasion* or *post-invasion* in nature. The objective of pre-invasion actions is to prevent non-native species from invading a new region. In contrast, post-invasion actions involve the optimal control of an alien species, given that this species has already invaded a new region. The purpose of our chapter is to shed light on the properties of one kind of pre-invasion action, namely, inspections. Hence, we now briefly review three papers on pre-invasion actions that are related to the analysis undertaken in our chapter.[52]

Horan *et al.* (2002) compare the properties of management strategies under full information and under uncertainty. Batabyal *et al.* (2005) show that, if reducing the economic cost associated with inspections is important, then it makes more sense for a seaport manager to select the inspection regime with fewer inspectors and less stringent inspections. In contrast, if diminishing the damage from biological invasions is more significant, then this manager ought to pick the inspection regime with more inspectors and more stringent inspections. Finally, Batabyal and Nijkamp (2005) have shown that, in an inspection cycle, irrespective of whether the inspection policy choice is made on the basis of an optimization exercise or on the basis of a rule of thumb, the "container policy" dominates the "temporal policy" because the container policy results in lower long run expected net cost (*LRENC*) from inspections. In this regard, the container policy involves the inspection of cargo upon the arrival of a specified number of containers in a seaport. In contrast, the temporal policy involves the inspection of container cargo at fixed points in time.

52For more on the literature on post-invasion actions, see Eiswerth and Johnson (2002), Eiswerth and van Kooten (2002), Olson and Roy (2002), and Acquaye *et al*. (2005).

Given the above finding, in this chapter, we continue the Batabyal and Nijkamp (2005) inspired line of inquiry and analyze container policies in three specific scenarios. In the first scenario, the time taken to conduct inspections is negligible. In the second scenario, this inspection time is deterministic, and in the third scenario, this inspection time is stochastic. Specifically, we compare and contrast the *LRENC* and the optimal container policy in each of these three scenarios.

The reader should note that whereas Batabyal and Nijkamp (2005) used a queuing theoretic model to study the "container versus time" question, in this chapter, we use the theory of renewal-reward processes to characterize optimal container policies in the three scenarios described in the previous paragraph.[53] The rest of this chapter is organized as follows. Section 2 first provides a primer on renewal-reward processes. Then, this section delineates a model of biological invasions in the context of a renewal-reward theoretic model of the container inspection policy choice problem faced by a *representative* inspector at a national point of entry such as an airport, a land border crossing, or a seaport. Section 3 analyzes the case in which inspections require virtually no time. Section 4 focuses on the case in which inspections require a deterministic amount of time. Section 5 analyzes

53 See Ross (1996, pp. 132-140) and Tijms (2003, pp. 39-50) for more on renewal-reward processes.

the case in which the inspection time is random. Section 6 concludes and offers suggestions for future research on the topic of invasive species management.

2. Renewal-Reward Processes and Biological Invasions

2.1. Preliminaries

As Ross (1996, pp. 98-162) has noted, a stochastic process $\{Z(t):t\geq 0\}$ is said to be a counting process if $Z(t)$ denotes the total number of counts that have taken place by time t. Clearly, since $Z(t\text{-}1)$, $Z(t)$, $Z(t+1)$, etc. are probabilistic, the time between any two counts $Z(t)$, $Z(t\text{-}1)$, is also probabilistic. This time between any two counts is called the interarrival time. A counting process in which the interarrival times have a *general* cumulative distribution function is a renewal process.

Consider a renewal process $\{Z(t):t\geq 0\}$ with interarrival times X_z $z\geq 1$, which have a cumulative distribution function denoted by $F(\bullet)$. In addition, suppose that a monetary reward R_z is earned when the *zth* renewal is completed. Let $R(t)$, the total reward earned by time t, be given by $\sum_{z=1}^{Z(t)} R_z$, and let $E[R_z]=E[R]$, and $E[X_z]=E[X]$. The renewal-reward theorem—see Ross (1996, p. 133) or Tijms (2003, p. 41)—tells us that if $E[R]$ and $E[X]$ are finite, then with probability one,

$$\lim_{t\to\infty} \frac{E[R(t)]}{t} = \frac{E[R]}{E[X]}. \tag{1}$$

In words, if we think of a cycle being completed every time a renewal occurs, then the long run average reward is simply the expected reward in a cycle divided by the expected time it takes to complete that cycle. With this background, we are now in a position to delineate our model of biological invasions and inspections.

2.2. A model of biological invasions

Consider a stylized national point of entry such as an airport, a land border crossing, or a seaport.[54] Upon arrival at this seaport, ships unload their containers carrying cargo. The arrival of these containers coincides with the arrival of a whole host of potentially injurious biological organisms. Now, before this incoming cargo can be moved to various points in the interior of the country under consideration, the containers must first be inspected. The purpose of this prophylactic activity is to decrease the likelihood of the occurrence of one or more biological invasions. We assume that the arrival rate of the various biological organisms is proportional to the arrival rate of the containers

54To fix ideas, in the rest of this chapter, we shall think of the national point of entry as a seaport. However, the reader should understand that our analysis is also applicable in the case in which the point of entry is either an airport or a land border crossing.

at the inspection facility. Therefore, we shall not model the biological organisms directly. Rather, we shall focus on the containers that bring these organisms to our seaport. Further, the arrival process of the containers at the inspection facility in our seaport represents the arrival process for the renewal-reward theoretic model that we analyze in this chapter. Finally, we suppose that the containers in question arrive at the inspection facility in accordance with a particular renewal process, namely, a Poisson process with rate λ and mean interarrival time $1/\lambda$.

Our seaport manager is interested in preventing invasions by the potentially deleterious biological organisms. Therefore, the cargo arriving in containers must be inspected. In this chapter, we analyze the inspection question from the standpoint of a representative inspector employed at the seaport under consideration. Now, cargo from different points of origin in the world will typically require different amounts of inspection time. For instance, if the port manager knows that a particular consignment of containers is from a country with which the manager's country has been trading for a long period of time without any invasions then there is relatively little risk of an invasion from this country. Alternately, if the country of origin has a cargo pre-certification program then, once again, the risk of invasions is low and therefore inspections at our seaport will require little or no time. To model this case, in section 3, we study the optimal container policy when inspections require a negligible amount of time.

In contrast to the above cases, when our seaport manager is confronted with container cargo from a country with a history of prior invasions or a country with which the manager's country has had limited trade in the past, inspections will not be mere formalities and hence will require non-trivial amounts of time. Finally, our seaport manager may be faced with container cargo from a relatively unfamiliar part of the world or this manager may be faced with container cargo that is not arriving directly on a ship from a particular country, but rather it does so indirectly, because the ship carrying the cargo has visited another seaport before arriving at our manager's seaport.[55] To model these two cases, in sections 5 and 6 we analyze the optimal container policy when inspections require, respectively, deterministic and stochastic amounts of time. We now formally analyze the optimal container policy for the case in which inspections require a negligible amount of time.

3. Negligible Inspection Times

Containers arrive at the inspection facility in our seaport in accordance with a Poisson process with rate λ and mean interarrival time $1/\lambda$. The port manager's inspector begins his job once N containers have arrived at the inspection facility. We assume that these inspections take a

55An example of this last case would be container cargo arriving in Vancouver, Canada, on a ship from Valparaiso, Chile, via Long Beach, in the United States.

negligible amount of time. We stress that the reader should not interpret this assumption literally. Instead, the reader should think of this case as one in which, because of the reasons given in the second paragraph of section 2.2, our inspector clears the consignment of N containers with virtually no delay.

As a result of these inspections, the inspector (and ultimately the port manager) incurs costs and obtains benefits from two sources. The first source of net cost, the difference between total cost and total benefit, arises from things like the expense of paying the inspector and holding the containers in the inspection facility (a cost) and from the non-contamination of the containers that are being held in the inspection facility (a benefit). We model this first source of net cost by supposing that our inspector incurs net cost at the rate of $h \bullet j$ dollars when j containers are waiting to be inspected. The second source of net cost arises from things like the negative impact on society from the slowdown in economic activity while the containers are being inspected (a cost) and from the determination of whether the containers actually contain what they are supposed to contain (a benefit). We account for this second source of net cost by supposing that the inspector incurs a fixed net cost of K dollars each time N containers have arrived and our inspector has begun his job of inspecting these N containers.

The inspection function is typically cyclical in nature with both busy and idle periods. Therefore, using the notation of section 2.1, if we let $Z(t)$ denote the number of containers at time t then the continuous time stochastic process $\{Z(t):t \geq 0\}$ probabilistically regenerates itself every time our inspector begins his job. This tells us that the expected length of an inspection cycle or E[*length of cycle*] equals N/λ. To compute the expected net cost per inspection cycle or E[*net cost per cycle*], note that mathematically these costs are given by

$$E[\textit{net cost per cycle}]=K+\frac{h(N-1)}{\lambda}+\frac{h(N-2)}{\lambda}$$

$$+\ldots+\frac{h}{\lambda}=K+\frac{hN(N-1)}{2\lambda}. \tag{2}$$

Now, recognizing that a cost is a negative reward, we can apply the renewal-reward theorem—see equation (1)—to our inspection cycle and thereby compute our port manager's *LRENC* from inspection activities. Specifically, the numerator of the ratio on the right-hand-side (RHS) of equation (1) is E[*net cost per cycle*] and the denominator is E[*length of cycle*]. Therefore, we have

$$LRENC=\frac{E[\textit{net cost per cycle}]}{E[\textit{length of cycle}]}=\frac{\lambda K}{N}+\frac{h(N-1)}{2}. \tag{3}$$

Having computed the *LRENC* of inspections, we are now in a position to state our seaport manager's optimization problem. Specifically, this manager chooses the number of containers N to minimize the *LRENC* from inspections. Formally, our port manager solves

$$\min_{\{N\}}\left[\frac{\lambda K}{N}+\frac{h(N-1)}{2}\right]. \tag{4}$$

Treating N as a continuous control variable and using calculus, we see that the optimal number of containers N_n^* that minimizes the port manager's *LRENC* is given by

$$N_n^*=\sqrt{\frac{2\lambda K}{h}}. \tag{5}$$

The value of N, N_n^* that minimizes the *LRENC* of inspections is given by equation (5). In words, the optimal number of containers N_n^* equals the square root of the ratio of the product of twice the Poisson arrival rate of containers (λ) and the second source of fixed net cost (K) to the first source of net cost parameter (h). Put differently, the optimal container policy when inspections take a negligible amount of time is to commence inspections when N_n^* containers have arrived at the inspection facility in our seaport.

Inspecting equation (5), it is easy to verify two properties of the optimal number of containers N_n^*. First, as the second and fixed source of net cost (*K*) increases, it is in the interest of the seaport manager to raise the optimal number of containers before commencing inspections. Second, if the first source of net cost parameter (*h*) goes up, then the seaport manager finds it desirable to lower the optimal number of containers before commencing inspections. Having characterized the port manager's *LRENC* (equation (3)) and the optimal container policy (equation (5)), we now move on to analyze the case in which inspections take a deterministic amount of time.

4. Deterministic Inspection Times

As in section 3, the cargo containers arrive in our seaport's inspection facility in accordance with a Poisson process with rate λ and mean interarrival time $1/\lambda$. However, in this section, for the reasons given in section 2.2, we suppose that inspections take *T* time periods to complete. To continue the analysis, let us first note that the stochastic process $\{Z(t):t\geq 0\}$ of section 3 is, once again, regenerative and this process regenerates itself when the inspection of *N* containers is complete. In this case, the computation of the expected length of an inspection cycle is straightforward and some thought tells us that $E[length\ of\ cycle]=N/\lambda+T$. To compute the expected net cost per cycle or

E[*net cost per cycle*], we use equation 1.1.10 in Tijms (2003, p. 17) and get

$$E[net\ cost\ per\ cycle] = K + \frac{hN(N-1)}{2\lambda} + hTN + \frac{h\lambda T^2}{2}. \tag{6}$$

We now use the renewal-reward theorem—equation (1)—and the two expectations that we have just derived and conclude that

$$LRENC = \frac{E[net\ cost\ per\ cycle]}{E[length\ of\ cycle]} =$$

$$\frac{\lambda K}{N+\lambda T} + \frac{hN(N-1)}{2(N+\lambda T)} + \frac{\lambda hNT}{N+\lambda T} + \frac{h\lambda^2 T^2}{2(N+\lambda T)}. \tag{7}$$

Comparing equations (3) and (7) we see that, consistent with our intuition, the *LRENC* when inspections take a deterministic and non-negligible amount of time is higher than the *LRENC* when inspections take a negligible amount of time. We can now state our seaport manager's optimization problem. Specifically, this manager selects the number of containers N to minimize the *LRENC* in equation (7). Formally, our port manager solves

$$\min_{\{N\}} \left[\frac{\lambda K}{N+\lambda T} + \frac{hN(N-1)}{2(N+\lambda T)} + \frac{\lambda hNT}{N+\lambda T} + \frac{h\lambda^2 T^2}{2(N+\lambda T)}\right]. \tag{8}$$

We now treat N as a continuous control variable, differentiate the minimand in equation (8) with respect to N, and then set the resulting expression equal to zero. After several steps of algebra, the problem in (8) can be reduced to determining the roots of a quadratic equation in N and that equation is

$$hN^2+2h\lambda TN+h(\lambda T)^2-2\lambda K-h\lambda T=0. \tag{9}$$

Using the standard method, the two possible roots of this quadratic equation are given by $\{-2h\lambda T\pm\sqrt{4h\lambda(2K+hT)}\}/2h$. Now, one of these roots—and therefore the optimal number of containers—is negative and hence we discard this root. This leaves us with the remaining root. Hence, the optimal number of containers N_d^* that minimizes the port manager's *LRENC* is given by

$$N_d^*=\frac{\sqrt{4h\lambda(2K+hT)}-2h\lambda T}{2h}. \tag{10}$$

In other words, the optimal container policy when inspections take a deterministic amount of time is to commence inspections when N_d^* containers have arrived at the inspection facility in our seaport. Inspecting equations (5) and (10) we see that the optimal container policies (N_n^* and N_d^*) in the negligible time and in the deterministic time cases are not directly comparable because of the relative complexity of the expression describing N_d^* in equation (10). Even so, it is straightforward to confirm that as $T \rightarrow 0$ we have $N_d^* \rightarrow N_n^*$. That is, as the inspection time approaches zero, the optimal (deterministic time) container policy N_d^* converges to the optimal (negligible time) container policy N_n^*.

Inspecting equation (10), we see that one comparative statics result from the negligible inspection time case analyzed in section 3 carries over to this section as well. Specifically, as the second and fixed source of net cost (K) increases, it is in the interest of the seaport manager to raise the optimal number of containers before commencing inspections. This completes our analysis of the optimal container policy when inspection times are deterministic. We now proceed to discuss the final scenario of this chapter. In this last scenario, inspections take a random amount of time.

5. Probabilistic Inspection Times

As in sections 3 and 4, the cargo containers arrive in our seaport's inspection facility in accordance with a Poisson process with rate λ and mean interarrival time $1/\lambda$. Following the reasons given in section 2.2, in this section, we suppose that inspections take a random amount of time to complete. To this end, let the time taken to complete inspections be a random variable with finite first two moments. Further, let us denote these two moments by $\mu_1 > 0$ and $\mu_2 > 0$ respectively.

Now, the stochastic process $\{Z(t): t \geq 0\}$ of sections 3 and 4 is, once again, regenerative and this process regenerates itself when the inspections of all N containers is complete. Because the random inspection times have finite first two moments, we can easily adapt the section 4 analysis to the scenario of this section. Specifically, the expected length of an inspection cycle or E[*length of cycle*] is now simply equal to $N/\lambda + \mu_1$. Further, to calculate the expected net cost per cycle or E[*net cost per cycle*] we replace hTN and $h\lambda T^2/2$ in equation (6) with $h\mu_1 N$ and $h\lambda\mu_2/2$ respectively. With these substitutions, we have

$$E[\textit{net cost per cycle}] = K + \frac{hN(N-1)}{2\lambda} + h\mu_1 N + \frac{h\lambda\mu_2}{2}. \qquad (11)$$

Now, using the renewal-reward theorem and the two expectations that we have just computed, we get

$$LRENC = \frac{E[net\ cost\ per\ cycle]}{E[length\ of\ cycle]} = \frac{\lambda K}{N+\lambda\mu_1} +$$

$$\frac{hN(N-1)}{2(N+\lambda\mu_1)} + \frac{\lambda hN\mu_1}{N+\lambda\mu_1} + \frac{h\lambda^2\mu_2}{2(N+\lambda\mu_1)}. \quad (12)$$

Comparing equations (3), (7), and (12) we immediately see two things. First, the *LRENC* with random inspection times is definitely larger than the *LRENC* with negligible inspection times. Second, whether the *LRENC* with random inspection times is larger or smaller than the *LRENC* with deterministic inspection times cannot be determined unambiguously. The general answer depends on the relative magnitudes of (T, μ_1) and (T^2, μ_2). To illustrate this point, we now highlight two special cases. First, if $\mu_1 = T$ and $\mu_2 = T^2$ then the two *LRENCs* in equations (7) and (12) are identical. Second, suppose $T = \mu_1$. Then, the *LRENC* with random inspection times is greater (lesser) than the *LRENC* with deterministic inspection times depending on whether $\mu_2 > (<) T^2$.

We now state our port manager's optimization problem. This manager selects the number of containers N to minimize the *LRENC* in equation (12). Formally, he solves

$$\min_{\{N\}}\left[\frac{\lambda K}{N+\lambda\mu_1}+\frac{hN(N-1)}{2(N+\lambda\mu_1)}+\frac{\lambda hN\mu_1}{N+\lambda\mu_1}+\frac{h\lambda^2\mu_2}{2(N+\lambda\mu_1)}\right]. \tag{13}$$

Following the procedure described in section 4, the problem in (13) can be reduced to determining the roots of a quadratic equation in N and that equation is

$$hN^2+2h\lambda\mu_1 N+h\lambda^2\mu_2-2\lambda K-h\lambda\mu_1=0. \tag{14}$$

Using the usual method, the two possible roots of the quadratic equation in (14) are given by $\{-2h\lambda\mu_1\pm\sqrt{4h\lambda(\lambda h\mu_1^2+2K+h\mu_1-\lambda h\mu_2)}\}/2h$. Now, if $\lambda h\mu_1^2+2K+h\mu_1\leq\lambda h\mu_2$ then the roots are either complex or negative and hence we suppose that $\lambda h\mu_1^2+2K+h\mu_1>\lambda h\mu_2$ holds. Even with this proviso, one of the two roots—and hence the optimal number of containers—is negative and therefore we eliminate this root. This leaves us with a single root and we conclude that the optimal number of containers N_p^* that minimizes the port manager's *LRENC* is given by

$$N_p^*=\frac{\sqrt{4h\lambda\left(\lambda h\mu_1^2+2K+h\mu_1-\lambda h\mu_2\right)}-2h\lambda\mu_1}{2h}. \tag{15}$$

Equation (15) tells us that the optimal container policy when inspections take a random amount of time to complete is to commence inspections when N_p^* containers have arrived at the inspection facility in our seaport. Inspecting equations (5) and (15), we see that the optimal container policies (N_n^* and N_p^*) in the negligible time and in the random time cases are not directly comparable because of the relative complexity of the N_p^* expression in equation (15). Even so, it is easy to verify that as $\mu_1 \rightarrow 0$ and $\mu_2 \rightarrow 0$ we have $N_p^* \rightarrow N_n^*$. That is, as the first and the second moments of the random inspection time approach zero, the optimal (random time) container policy N_p^* converges to the optimal (negligible time) container policy N_n^*. Finally, comparing equations (10) and (15) we see that when $T = \mu_1$ and $\mu_2 = \mu_1^2$, the optimal container policies for the deterministic time and the random time inspection scenarios are identical.

Examining equation (15), we see that as in section 4, one comparative statics result for the negligible inspection time scenario analyzed in section 3 carries over to this section as well. In particular, as the second and fixed source of net cost (K) increases, the seaport manager will want to raise the optimal number of containers, before commencing inspections. This completes our characterization of the port manager's *LRENC* (equation (12)) and his optimal container policy (equation (15)) when inspections take a random amount of time to complete.

6. Conclusions

In this chapter, we continued a line of inquiry begun recently by Batabyal and Nijkamp (2005). Specifically, we analyzed container policies in three scenarios. In the first scenario, the time taken to conduct inspections is negligible. In the second scenario, this inspection time is deterministic and finally in the third scenario this inspection time is probabilistic. We compared and contrasted the *LRENC* and the optimal container policy in each of these three scenarios. Our analysis shows that, *inter alia*, the *LRENC* when inspection times are either deterministic or probabilistic are typically higher than the corresponding *LRENC* when inspection times are negligible. In addition, although the *LRENC* for the deterministic and the probabilistic inspection time cases are identical in a special case, in general, these two cost expressions are dissimilar.

The analysis in this chapter can be extended in a number of directions. In what follows, we make two suggestions for extending the research described in this chapter. First, it would be useful to analyze scenarios in which the inspection times within a particular inspection regime are container specific. Second, it would also be helpful to ascertain whether the results we have obtained in this chapter with the Poisson process carry over to more general renewal processes. Studies of inspections in invasive species management that incorporate these aspects of the problem into the analysis will provide additional insights into a management task that has considerable economic and biological ramifications.

References

Acquaye, A.K.A., Alston, J.M., Lee, H., and Sumner, D.A. 2005. Economic consequences of invasive species policies in the presence of commodity programs: Theory and application to Citrus Canker, *Review of Agricultural Economics*, 27, 498-504.

Batabyal, A.A. 2004. A research agenda for the study of the regulation of invasive species introduced unintentionally via maritime trade, *Journal of Economic Research*, 9, 191-216.

Batabyal, A.A., and Nijkamp, P. 2005. On container versus time based inspection policies in invasive species management, *Stochastic Environmental Research and Risk Assessment*, 19, 340-347.

Batabyal, A.A., Beladi, H., and Koo, W.W. 2005. Maritime trade, biological invasions, and the properties of alternate inspection regimes, *Stochastic Environmental Research and Risk Assessment*, 19, 184-190.

Batabyal, A.A., and Beladi, H. 2006. International trade and biological invasions: A queuing theoretic analysis of the prevention problem, *European Journal of Operational Research*, 170, 758-770.

Cox, G.W. 1993. *Conservation Ecology*. W.C. Brown Publishers, Dubuque, IA.

Eiswerth, M.E., and Johnson, W.S. 2002. Managing nonindigenous invasive species: Insights from dynamic analysis, *Environmental and Resource Economics*, 23, 319-342.

Eiswerth, M.E., and van Kooten, G.C. 2002. Uncertainty, economics, and the spread of an invasive plant species, *American Journal of Agricultural Economics*, 84, 1317-1322.

Feenstra, R.C. 2004. *Advanced International Trade*. Princeton University Press, Princeton, NJ.

Horan, R.D., Perrings, C., Lupi, F., and Bulte, E.H. 2002. Biological pollution prevention strategies under ignorance: The case of invasive species, *American Journal of Agricultural Economics*, 84, 1303-1310.

Krugman, P.R., and Obstfeld, M. 2005. *International Economics*, 7th edition. Pearson Addison Wesley, Boston, MA.

Office of Technology Assessment (OTA). 1993. *Harmful Non-Indigenous Species in the United States*. OTA-F-565, Washington, DC.

Olson, L.J., and Roy, S. 2002. The economics of controlling a stochastic biological invasion, *American Journal of Agricultural Economics*, 84, 1311-1316.

Parker, I.M., Simberloff, D., Lonsdale, W.M., Goodell, K., Wonham, M., Kareiva, P.M., Williamson, M.H., Von Holle, B., Moyle, P.B., Byers, J.E., and Goldwasser, L. 1999. Impact: Toward a framework for understanding the ecological effects of invaders, *Biological Invasions*, 1, 3-19.

Pimentel, D., Lach, L., Zuniga, R., and Morrison, D. 2000. Environmental and economic costs of nonindigenous species in the United States, *BioScience*, 50, 53-65.

Ross, S.M. 1996. *Stochastic Processes*, 2nd edition. Wiley, New York, NY.

Tijms, H.C. 2003. *A First Course in Stochastic Models*. Wiley, Chichester, UK.

USDA, APHIS, and USFS. 2000. Pest Risk Assessment for Importation of Solid Wood Packing Materials into the United States. http://www.aphis.usda.gov/ppq/pra/swpm/complete.pdf

Vitousek, P.M., D'Antonio, C.M., Loope, L.L., and Westbrooks, R. 1996. Biological invasions as global environmental change, *American Scientist*, 84, 468-478.

Work, T.T., McCullough, D.G., Cavey, J.F., and Komsa, R. 2005. Arrival rate of nonindigenous insect species into the United States through foreign trade, *Biological Invasions*, 7, 323-332.

Yang, Z., and Perakis, A.N. 2004. Multiattribute decision analysis of mandatory ballast water treatment measures in the US great lakes, *Transportation Research Part D*, 9, 81-86

CHAPTER 11

AN ANALYSIS OF ECONOMIC COST MINIMIZATION AND BIOLOGICAL INVASION DAMAGE CONTROL USING THE *AWQ* CRITERION - WITH G.J. DEANGELO AND S. KUMAR

Abstract. DeAngelo *et al.* (2007) have recently used the *AWS* criterion in a *M/G/*1 queuing model to show that there is *no* necessary tension between economic cost minimization and inspection stringency in non-native species management. In this chapter, we use an alternate cost criterion (*AWQ* criterion) to investigate the generality of this central result in DeAngelo *et al.* (2007). Our theoretical analysis shows that there is no unambiguous answer to this question. Therefore, we use numerical methods, and our numerical analysis leads to two findings. First, for many values of the model parameters that describe the strictness of inspections, there *is* a tension between cost minimization and inspection stringency. Second, for most values of the model parameter that depicts the volume of maritime trade handled by the seaport under consideration, there is *no* tension between cost minimization and inspection stringency.

1. Introduction

In modern times, airplanes, ships, and trucks have all been used to carry goods from one region of the world to another. Representative textbooks in international trade such as Feenstra (2004) and Krugman and Obstfeld (2005) tell us that unrestricted trade between different regions of the world is welfare improving. This notwithstanding, researchers in both life sciences and the social sciences have increasingly pointed out that the extent of these welfare improvements is likely to be less than what most investigators have hitherto posited. Why? As Parker *et al.* (1999), Batabyal (2004), and Work *et al.* (2005) have noted, this is because, in addition to carrying goods between regions, airplanes, ships, and trucks have also managed to carry a whole host of non-native plant and animal species (also known as alien or invasive species) from one geographical region of the world to another.

There are many ways in which airplanes, ships, and trucks have transported non-native plant and animal species from one region of the world to another. Non-native animal species have sometimes succeeded in lodging themselves in the landing gear of airplanes and in this way they have traveled as stowaways from one nation to another. Similarly, a number of marine non-native species have been introduced unintentionally into a region by ships dumping their ballast water. Cargo ships commonly carry ballast water in order to augment vessel stability when they are not carrying full loads. When these ships come into a seaport, this ballast water must be released before cargo can be loaded. This means of species introductions is important and, very recently, the problem of managing alien species that have been introduced into a particular region by means of the dumping of ballast water has received some attention in the literature.[56]

Ships routinely use containers to carry cargo from one nation to another and these containers are often the source for the introduction of one or more non-native species. Such introductions take place because non-native species can remain hidden in containers for long periods of time. In addition, substances, such as wood—often used to pack cargo in containers—may themselves contain non-native species. In fact, as noted by Batabyal and Nijkamp (2005), a joint report from the United States Department of Agriculture (USDA), the Animal and Plant Health Inspection Service (APHIS), and the United States Forest Service (USFS) has shown that nearly 51.8% of maritime shipments contain solid wood packing substances and that infection rates for solid wood packing substances are non-trivial (USDA, APHIS, and USFS (2000, p. 25)). To see this, consider the following case. Inspections of wooden spools from China revealed infection rates between 22% and 24% and

56See Yang and Perakis (2004), Batabyal *et al.* (2005), and Batabyal and Beladi (2006) for more on this literature.

inspections of braces for granite blocks imported into Canada were found to hold live insects 32% of the time (USDA, APHIS, and USFS (2000, pp. 27-28)).

When non-native species invade new habitats, they give rise to immense costs in the nations in which these novel habitats are located. Here are two illustrations of such costs from the United States. First, the Office of Technology Assessment (OTA (1993)) has established that the Russian wheat aphid caused $600 million worth of crop damage between 1987 and 1989. Second, Pimentel *et al.* (2000) have calculated the total costs of all non-native species at around $137 billion per year. Economic costs are not the end of the story. In fact, in addition to these economic costs, non-native species have caused a lot of ecological damage as well. For example, Vitousek *et al.* (1996) have noted that non-native species can change ecosystem processes, act as vectors of diseases, and diminish biological diversity. In addition, Cox (1993) has observed that out of 256 vertebrate extinctions with a known cause, 109 are the outcome of the actions of non-native species. Because of these reasons, both economists and ecologists are very interested in the management of non-native species.

Now, from the standpoint of a manager, there are a number of actions that this individual can take to address the problem of biological invasions. These actions are typically *pre-invasion* or *post-invasion* in nature. The objective of pre-invasion actions is to prevent non-native species from invading a new habitat. In contrast, post-invasion actions involve the optimal control of a non-native species, given that this species has already invaded a new habitat. The nascent economics literature on the management of non-native species has generally focused its attention on the properties of alternate post-invasion actions. For instance, Barbier (2001) has noted that the economic impact of a biological invasion can be discerned by analyzing the nature of the interaction between the native and the alien species. Using an intertemporal management model, Eiswerth and Johnson (2002) point out that the optimal level of management effort is responsive to ecological factors that are not only species and site specific, but also probabilistic in nature. Olson and Roy (2002) have used a stochastic framework to examine the situations in which it is optimal to wipe out a non-native species and the situations in which it is not optimal to do so. Finally, Eiswerth and van Kooten (2002) have shown that, in some cases, it is possible to use information furnished by specialists to create a model in which it is optimal to not eradicate but instead regulate the spread of an alien species.

Inspections are a basic pre-invasion tool that is available to managers interested in precluding biological invasions. They are routinely used at airports, land border crossings, and in seaports to screen humans, the cargo carried by humans, and the cargo carried in containers. As a result, several researchers have now begun to formally study inspections and their properties in the context of non-native species management. In this regard, McAusland and Costello (2004), Batabyal *et al.* (2005), Batabyal and Beladi (2006), and DeAngelo *et al.* (2007) have all studied the properties of alternate ways of structuring the inspection function given that inspections are a very useful practical pre-invasion management tool. McAusland and Costello (2004) show that, when one takes a dynamic view and considers the future effects of current species introductions, one is naturally led to favor more stringent inspections. Batabyal and Beladi (2006) use queuing theory to show how maximization problems for choosing the optimal number of inspectors can be formulated after one undertakes a stationary state analysis of two multi-person inspection regimes. Batabyal *et al.* (2005) have analyzed *M/M/*1 and *M/M/*2 queuing models of inspections and have concluded that there *is* a tension between economic cost minimization and inspection stringency in non-native species management. What this means is that greater inspection stringency with a larger number of inspectors leads to higher economic costs and laxer inspection stringency with a smaller number of inspectors results in lower economic costs.

The paper that is most closely related to our chapter is the one by DeAngelo *et al.* (2007). Focusing their attention on an arbitrary seaport, these researchers have studied the properties of inspections by utilizing the "average wait of a ship in the port system" or *AWS* criterion in a *M/G/*1 queuing model. Their basic result is that there is *no* necessary tension between economic cost minimization and inspection stringency in non-native species management. In the context of their paper, greater (lesser) inspection stringency reflects an enhanced (decreased) concern for the potential damage from a biological invasion. Therefore, a seaport manager who places a relatively big (small) weight on invasion damage control will, all else being equal, want to inspect ships more (less) stringently.

As we have noted previously, inspections clearly constitute an important part of the general task of pre-invasion non-native species management. Therefore, we use an alternate cost criterion, namely the "average wait of a ship in queue" or *AWQ* criterion to investigate the generality of the central "no necessary tension" result in DeAngelo *et al.* (2007). Note that we are not saying that *AWS* is a useless measure of costs. On the contrary, it is a perfectly reasonable way to model the economic cost of inspections. Further, we stress that, within the context of a queuing model of non-native species management, there are two standard ways of modeling the economic cost imposed on society by inspections.[57] DeAngelo *et al.* (2007) have used *AWS* and have hence

57For more on this point, see Ross (2003, p. 477).

concentrated on the first of these two ways. Therefore, a natural question to ask is whether the central DeAngelo *et al.* (2007) result holds when costs are modeled using the second of these two ways. That is what we're doing in this chapter by focusing on *AWQ*, and this is the sense in which our chapter is an extension of the earlier DeAngelo *et al.* (2007) paper.

Our theoretical analysis demonstrates that there is an ambiguous answer to this tension question. Hence, we use numerical methods, and our numerical analysis leads to two conclusions. First, for many values of the model parameters that describe the strictness of inspections, there *is* a tension between cost minimization and inspection stringency. Second, for most values of the model parameter that depicts the volume of maritime trade handled by the seaport under consideration, there is *no* tension between cost minimization and inspection stringency. Our results in this chapter are different from the findings in Batabyal *et al.* (2005) with the *M/M/*1 and the *M/M/*2 queuing models. The general reason for this variance is that the queuing model that we employ in this chapter is *different* from the queuing models used by Batabyal *et al.* (2005). More specifically, the random inspection times in the Batabyal *et al.* (2005) paper are *exponentially* distributed, and the exponential distribution is the unique distribution that has the so called memoryless property.[58] In contrast, in the present chapter, the inspection times are *generally,* and not exponentially, distributed.

The rest of this chapter is arranged as follows. Section 2 provides a primer on queuing theory, and then this section describes the queuing theoretic model that we use in this chapter. Section 3 first delineates the "average wait of a ship in queue" or *AWQ* criterion that we use in this chapter to explore the generality of the findings obtained by DeAngelo *et al.* (2007). Next, this third section conducts a detailed theoretical and numerical analysis of the aforementioned tension question. Section 4 concludes and offers suggestions for future research on the subject of this chapter.

2. Economic Cost Minimization and Inspection Stringency

2.1. Queuing theory: a primer

Queuing theory—see Taylor and Karlin (1998) and Ross (2003) for textbook expositions—studies waiting lines or queues from a mathematical perspective. All queuing models have the following three features. First, there is a random arrival process. Second, there is a probabilistic service process. Finally, there is a deterministic number of

58For additional details on this property, see Ross (2003, pp. 272-273).

servers. In the queuing model of this chapter, the Poisson process constitutes the arrival process. In this case, the times between successive arrivals are exponentially distributed and the exponential distribution is memoryless. Therefore, the letter *M* is commonly used to describe the Poisson arrival process.[59]

Like the interarrival times of the previous paragraph, in general, the service or inspection times are also stochastic. The reader will recall that the main aim of our chapter is to investigate the generality of the findings in DeAngelo *et al.* (2007). Therefore, like DeAngelo *et al.* (2007), we also use the letter *G* to denote the *general* cumulative distribution function of the random inspection times.[60] Finally, the deterministic number of inspectors is generally denoted by some positive integer, and, in our chapter, we suppose that this positive integer is one as well. In the language of queuing theory, the inspection regime that we're analyzing in this chapter corresponds to the *M/G/*1 queuing model. In words, in our model, the arrival process of ships in the seaport is

59We use the Poisson process in part because this process has been used extensively in the literature on natural resources and the environment to model arrivals. For a more detailed corroboration of this claim, see Uhler and Bradley (1970), Arrow and Chang (1980), Mangel (1985), and Batabyal (2003).

60Note that when we say that the random inspection time is arbitrarily distributed, this does not mean that there are absolutely no restrictions imposed on this random variable. In fact, we are restricting the support of this random variable to be the non-negative real line. A second restriction is that we need the inspection time random variable's first two moments (mean and variance) to be finite. Beyond these two basic restrictions, we intend our analysis in this chapter to be as general as possible. That is why, in what follows, we begin with the theoretical analysis and then proceed to the numerical analysis when it is clear that the theoretical analysis will not provide unambiguous results. Finally, it is quite possible that the idiosyncracies of inspections in specific practical situations will impose further restrictions on the level of inspection stringency that a seaport manager will want to have in place. We suppose that these restrictions are captured in different values of the first two moments of the inspection time random variable.

Poisson, the time it takes to inspect a ship is arbitrarily or generally distributed, and there is one representative inspector.

2.2. Inspections in alien species management

Consider a stylized seaport in a specific coastal region of some nation that is publically owned. Ships with ballast water and/or cargo in containers arrive at this seaport to either load or unload cargo. If they have arrived to load cargo, then these ships carry this cargo to a seaport in some other region of the world. The arrival of these ships coincides with the arrival of a whole host of possibly damaging non-native animal and plant species. We suppose that the arrival rate of these animal and plant species is proportional to the arrival rate of the ships and, hence, we shall not model these species directly. Instead, we shall concentrate on the ships that bring these species to our seaport by means of either their ballast water or the containers that are used to carry the cargo. The arrival process of the ships in our seaport denotes the arrival process for the queuing theoretic inspection regimes that we analyze in this chapter.

Following the discussion in section 2.1, we assume that the ships in question arrive at our seaport in conformity with a Poisson process with rate $\lambda > 0$. The reader should note that, *ceteris paribus*, a higher λ indicates two things. First, because the arrival rate of the various non-native animal and plant species is proportional to the arrival rate of the ships, a higher λ means a larger volume of possibly deleterious biological organisms and hence a higher chance of a biological invasion. Second, a higher λ also means that our seaport is now dealing with more cargo or a higher volume of maritime trade. This discussion tells us that λ serves as a proxy for both the likelihood of a biological invasion and the volume of maritime trade.

A key objective of our seaport manager is to prevent invasions by the potentially destructive animal and plant species entering the seaport under study. Hence, arriving ships must be inspected before they can either load or unload cargo. We suppose that ships are inspected on a first-come-first-served basis and one inspector is assigned to each dock in our seaport. In what follows, we shall analyze a *representative* dock inspector's decision problem. Also, we shall think of the inspection function broadly. What this means is that for some ships, only the ballast water will be inspected. For other ships, only the containers carrying cargo will require inspection. Finally, for a third group of ships, both the ballast water and the containers will be inspected. Therefore, inspections will generally require random amounts of time. To model this explicitly, we allow the inspection times to be arbitrarily distributed. The seaport system under study consists of ships that are being inspected, ships that are waiting to be inspected, the representative dock inspector, and the seaport manager.

It is reasonable to expect that the amount of time it takes to complete inspections has a direct bearing on the stringency of inspections. Hence, to model this idea, we assume that there are two feasible inspection regimes in our seaport. In the first, or *more* stringent inspection regime, the average inspection time is v_M and the variance of this time is τ_M^2. In the second, or *less* stringent inspection regime, the mean inspection time is v_L and the variance of this time is τ_L^2. Further, we assume that $v_M > v_L$ and that $\tau_M^2 < \tau_L^2$ These two inequalities together tell us that inspection regime *M* is *more* stringent than inspection regime *L*. Why? Because, relative to regime *L*, on average, regime *M* requires that more time be spent on the inspection function. Also, the variability of the time spent inspecting ships in regime *M* is less than the variability of the time spent inspecting ships in regime *L*. As a consequence, the more stringent *M* regime's inspection times have a smaller coefficient of variation and thus are consistently smaller compared to those of the less stringent *L* regime.

The reader should note that we are measuring the *reliability* of the two inspection regimes with the two *variance* parameters τ_M^2 and τ_L^2. In other words, the stringent regime is stringent, not only because on average the inspector spends more time inspecting ships, but also because, having spent more time, the inspector can be *more certain* that he has not made either type I or type II errors, while discharging his duties. This feeling "more certain" corresponds to greater reliability, and this greater reliability, in turn, corresponds to a lower variance. That is why we have $\tau_M^2 < \tau_L^2$.[61] Given this interpretation, note that the $\tau_M^2 > \tau_L^2$ case is intuitively and practically implausible because this case would mean that, even though the inspector spends more time on average in the stringent regime inspecting ships, he is still less certain about the reliability of the inspection that he has undertaken.

The basic description of our two queuing theoretic inspection regimes is now complete. The more stringent inspection regime M corresponds to a *M/G/*1 queuing model with average inspection time v_M and variance τ_M^2 and the less stringent or L inspection regime also corresponds to a *M/G/*1 queuing model but now with mean inspection time v_L and variance τ_L^2. The reader should understand the manner in which we have mathematically depicted the basic question of this chapter: When seeking to preclude a biological invasion by inspecting the ballast water and/or the containers of ships, which inspection regime, *M* or *L,* ought our seaport manager to have in place? We now proceed to

61 There is support in the extant literature for our "lower (higher) variance implies more (less) reliable" interpretation. For additional details on this point, see Salvucci *et al.* (1997) and Yu (2005).

theoretically and numerically analyze the inspection regime choice question for the *AWQ* cost criterion that we identified in section 1.

3. The *AWQ* criterion

Inspections that lead to the preclusion of a biological invasion by non-native animal or plant species undoubtedly result in benefits to the citizens of the coastal region that we are studying. However, during the time that arriving ships are being inspected, there is neither loading nor unloading of cargo, and hence, in general, economic activity resulting from maritime trade is at a standstill. This ephemeral stoppage of economic activities imposes *costs* on the economy of our coastal region. This cost can be measured by computing a specific criterion, such as "the average wait of a ship in the port system" or *AWS*. In this way of looking at the problem, the longer (shorter) the average wait in the port system or *AWS* the larger (smaller) the costs from the cessation of economic activities. Therefore, a seaport manager who is concerned primarily about the economic costs that are imposed on society by the activities of the representative inspector will want to keep *AWS* as low as possible. In contrast, a seaport manager who frets more about the possible damage to society from a biological invasion will want to have the more stringent or *M* inspection regime in place. DeAngelo *et al.* (2007) have used this *AWS* criterion in a *M/G/*1 queuing model and have shown that contrary to the central finding in Batabyal *et al.* (2005), there is *no* universal tension between economic cost minimization and inspection stringency in non-native species management.

As noted in section 1, the *AWS* criterion is one of two possible criteria that we can use in thinking about the above mentioned economic costs, in the context of a queuing model. As such, suppose we adopt a somewhat looser interpretation of these costs and say that the loading and/or the unloading of cargo may proceed on a ship that is presently being inspected for potentially invasive species, but that such activities may not take place on ships that have yet to be inspected and are waiting in queue. In this way of looking at the problem, ships in *queue* are the ones that are most likely to impose economic costs on society. We shall refer to this second—and somewhat looser—economic cost criterion as the "average wait of a ship in queue" or *AWQ* criterion. Given this alternate cost criterion, a key issue that arises is expressed by the following question: Do the findings contained in DeAngelo *et al.* (2007) hold when the *AWQ* criterion and not the *AWS* criterion is used to measure the economic cost of inspections? We now address this question in detail.

Our first order of business is to compute *AWQ* for the two *M/G/*1 inspection regimes that we are presently studying. To do this, we shall use equations 1.2 and 3.17 in Taylor and Karlin (1998, p. 544 and p. 563). Using these two equations, our desired expressions for *AWQ* are

$$AWQ_M = \{\lambda(\tau_M^2 + v_M^2)\}/\{2(1-\lambda v_M)\} \text{ and } AWQ_L = \{\lambda(\tau_L^2 + v_L^2)\}/\{2(1-\lambda v_L)\} \quad (1)$$

respectively for the two regimes. We can use the inequalities $v_M > v_L$ and $\tau_M^2 < \tau_L^2$ to show that $2(1-\lambda v_M) < 2(1-\lambda v_L)$. However, because $(\tau_M^2 + v_M^2)$ may be larger or smaller than $(\tau_L^2 + v_L^2)$, simply knowing that $v_M > v_L$ and that $\tau_M^2 < \tau_L^2$ does *not* permit us to say anything unambiguous about the relative magnitudes of AWQ_M and AWQ_L. In other words, when our seaport manager sanctions the use of the more stringent M inspection regime in the seaport under study, it is *not* always the case that economic costs measured by the *AWQ* criterion will be higher. So, in the general case, there may or may not be a tension between economic cost minimization and inspection stringency. This tells us that, even when we use the *AWQ* criterion, the finding in DeAngelo *et al.* (2007) that there is no necessary tension between economic cost minimization and inspection stringency holds.

We have just seen that there is no unique resolution of this tension question. To provide additional insight into this issue, in the remainder of this chapter, we use a numerical perspective to study this tension question in three different ways. However, before we move to the specifics of the numerical analysis, let us stress three points. First, it is clearly infeasible to work with the infinitely many combinations of the model parameters $(\lambda, v_M, v_L, \tau_M^2, \tau_L^2)$ that are possible. As such, the reader should understand that the primary point of the subsequent numerical analysis is *illustrative*. We wish to demonstrate not only the functional dependence of the *AWQ* criterion on the various model parameters but also the *different* results that we obtain as we *vary* these parameters. Second, we have conducted numerical analyses of the sort delineated below, with many different values of the distinct model parameters. Rather than bore the reader with umpteen graphs of the various possible results, what we are reporting below is a selection of model parameters that generates—to the best of our knowledge—*representative* results. Finally, the collective lesson from the above two points is that the answer to the central tension question that we are investigating in this chapter is *seaport specific*. Different seaports are likely to have different values of the arrival rate λ. Therefore, in a particular practical situation, a seaport manager will typically want to choose the inspection regime stringency parameters (v, τ) to account for the specific value of λ confronting him as he optimizes a particular objective function such as *AWQ*.[62]

Let the arrival rate of ships be $\lambda=1$ per unit time. Further, suppose that the parameters of the two inspection regimes are $(v_M, \tau_M^2$

62 Clearly, *AWQ* is not the only possible objective function for a seaport manager. Depending on the specifics of a particular situation, a seaport manager may want to minimize costs in addition to those that we are capturing in *AWQ*.

)=(0.5,0.2) and (v_L, τ_L^2)=(0.4,0.9). Now, employing equation (1), it is straightforward to confirm that AWQ_M=0.45λ/(2-λ) and AWQ_L =1.06λ/(2-0.8λ). When λ=1, these two expressions reduce to AWQ_M =0.45 and AWQ_L=0.88.

Examining these two expressions for AWQ_M (AWQ_L), we see that, as the arrival rate of ships λ approaches 2 (2.5), economic costs measured by the AWQ_M (AWQ_L) criterion approach infinity. This means that there *is* an upper limit on the volume of maritime trade that our seaport can handle and, when this limit is approached, the economic costs of inspections become immeasurably large. Second, when v_M, v_L, τ_M^2, and τ_L^2 are as stated in the previous paragraph, the economic costs of inspections are lower when the more stringent M inspection regime is in place. The reader will note that this is also a case in which there is *no* tension between economic cost reduction and inspection stringency.

3.1. The tension question in terms of the volume of maritime trade

For the parameter values specified above, we now analyze the dependence of AWQ on λ. Equating AWQ_M =0.45λ/(2-λ)=0.45 and AWQ_L =1.06λ/(2-0.8λ)=0.88 and then simplifying gives us

$$0.70\lambda^2-1.22\lambda=0. \tag{2}$$

The two solutions to equation (2) are λ_1^*=1.74 and λ_2^*=0. Because λ must be positive, we conclude that λ=1.74 is the only economically meaningful solution in this case. Figure 1 plots AWQ on the vertical axis against the arrival rate of ships λ on the horizontal axis.

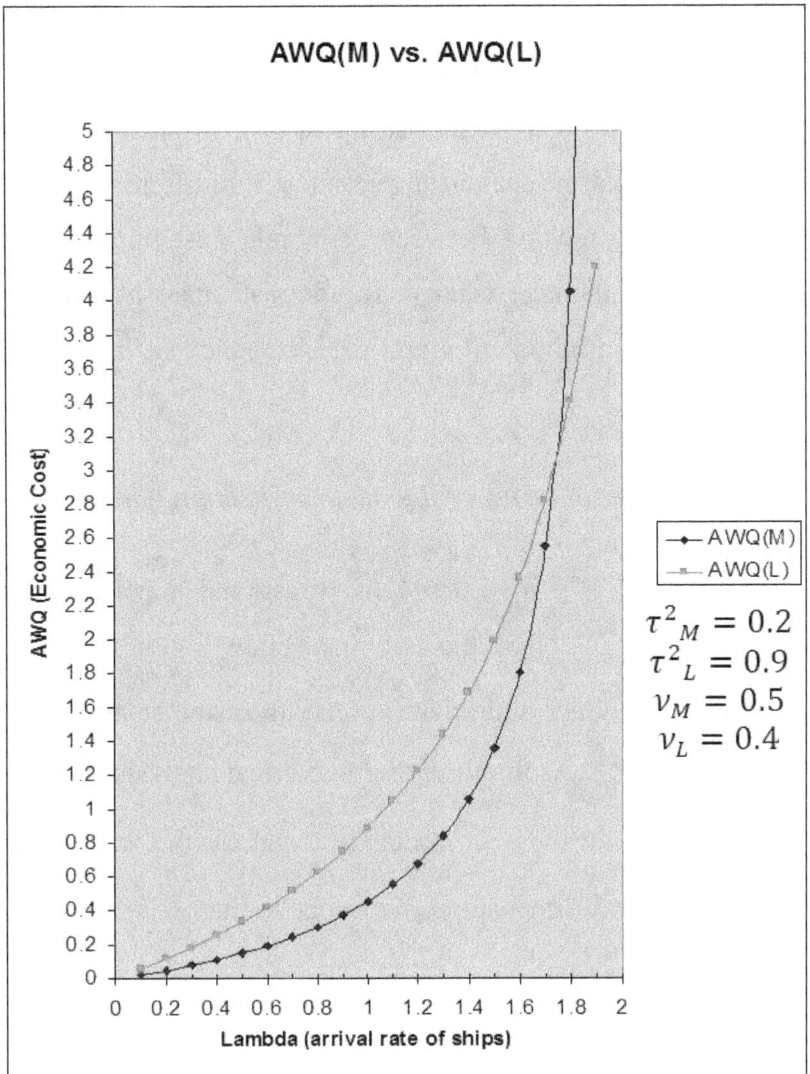

Figure 1

Looking at figure 1, we see that, for all values of λ, $0<\lambda\leq 1.74$, our seaport manager will prefer to have the more stringent or *M* inspection regime in place, because this regime leads to lower economic costs as measured by the *AWQ* criterion. Only when $\lambda > 1.74$, does it make sense for the seaport manager to use the less stringent or *L* inspection regime to inspect arriving ships at the seaport under study. Put differently, when $\lambda \in (0,1.74]$,, there is *no* tension between economic cost minimization and biological invasion damage control. In contrast, there *is* a tension between economic cost minimization and the stringency of inspections when $\lambda > 1.74$.

3.2. The tension question in terms of the mean inspection times

We now numerically investigate the functional dependence of *AWQ* on the means (v_M, v_L) of the two inspection regimes, and, in section 3.3, we shall conduct a similar exercise from the standpoint of the two variances (τ_M^2, τ_L^2). As in our earlier numerical analysis, we have $v_M > v_L$ and $\tau_M^2 < \tau_L^2$. Further, in order to conduct the subsequent numerical analysis in two dimensions, we suppose that $v_M = av_L$, $a > 1$ and that $\tau_M^2 = \mathrm{b}\tau_L^2$, $b \in (0,1)$. In other words, the two means and the two variances are assumed to be linearly related to each other and the parameters *a* and *b* are the two constants of proportionality. From an economic standpoint, we can think of the parameter *a* as a measure of the difference in the stringencies of the two inspection regimes *M* and *L*. Similarly, the parameter *b* can be thought of as a measure of the difference in the variability—and hence the reliability—of the same two inspection regimes.

Now, using the section 3 parameter values, we have λ=1, v_M= av_L, v_L=0.4, $\tau_M^2 = b\tau_L^2$, τ_L^2=0.9 and we set b equal to its midpoint, i.e., b=0.5. Note that this b=0.5 stipulation means that the more stringent inspection regime M is twice as reliable as the less stringent regime L. Using these values of the different parameters in equation (1), we get AWQ_M=(0.45+0.16a^2)/(2-0.8a) and AWQ_L=0.8833. Setting these two values equal gives us the following quadratic equation in a

$$0.16a^2+0.71a-1.32=0. \quad (3)$$

The two solutions to equation (3) are a_1^*=1.41 and a_2^*= -.5.83. Because we must have a>1, it follows that the only economically meaningful solution to equation (3) is a_1^*=1.41.

Figure 2 plots the economic cost criterion AWQ on the vertical axis against alternate values of a on the horizontal axis. Looking at figure 2 we see that when a=1.41, our seaport manager is indifferent between the two inspection regimes. Further, for all a<1.41 the use of the more stringent M inspection regime results in lower economic costs as measured by the AWQ criterion. Finally, for all $a > 1.41$, the use of the less stringent L inspection regime leads to lower economic costs.

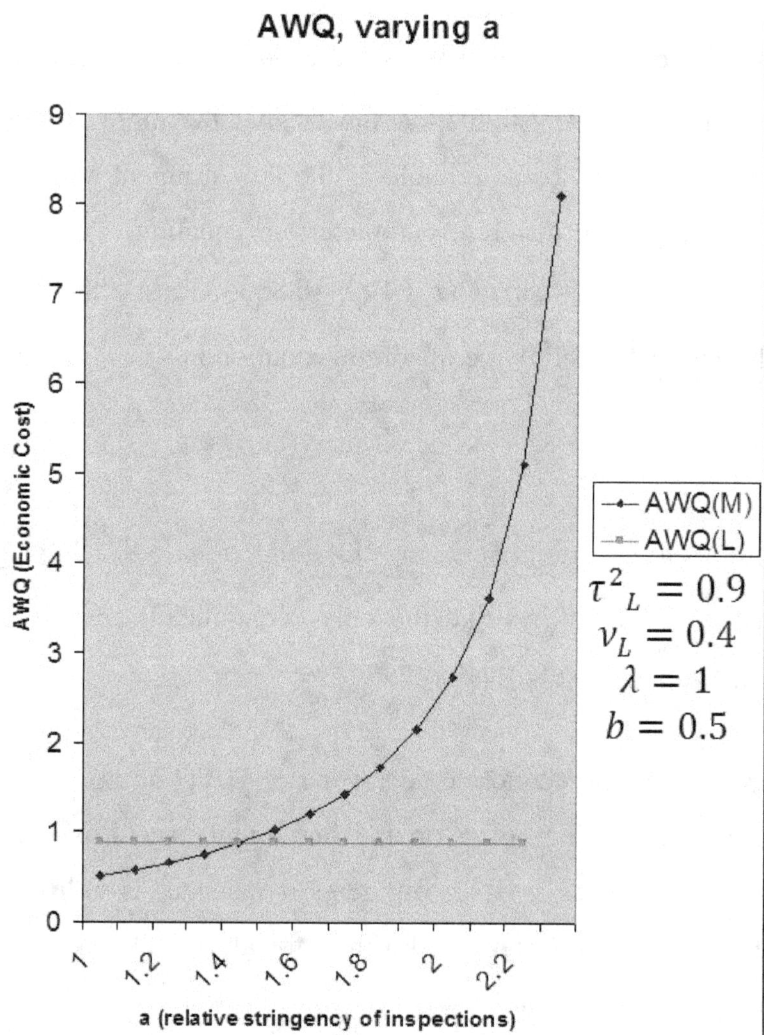

Figure 2

This tells us that, when $a>1.41$, there *is* a tension between economic cost minimization and biological invasion damage control.

In contrast, when a lies in the interval (1,1.41], there is *no* tension between economic cost minimization and biological invasion damage control. Note that this last "no tension" result may appear counterintuitive, but it holds for some and *not* all values of a. In addition, this last result does depend on our assumption that $\tau_M^2 < \tau_L^2$, but as we have already pointed out in section 2.2, the alternate assumption that $\tau_M^2 > \tau_L^2$ is both intuitively and practically improbable.

3.3. The tension question in terms of the variances of the inspection times

Our final task is to numerically examine the functional dependence of AWQ on the variances (τ_M^2, τ_L^2) of the M and the L inspection regimes. As in section 3.2, in order to conduct the analysis in two dimensions, we have $\tau_M^2 < \tau_L^2$, $v_M > v_L$, $\tau_M^2 = \mathrm{b}\tau_L^2$, $b \in (0,1)$, and $v_M = av_L$, $a>1$. Further, using the previous values of the pertinent parameters, we have λ=1, v_L=0.4, τ_L^2,=0.9, and we shall set a=2. The reader should note that setting a=2 means that, relative to the less stringent inspection regime *L,* on average, twice as much time is spent in the more stringent regime *M*. Substituting these values of the various parameters in equation (1), we get AWQ_M=1.6+2.25b and AWQ_L

=0.8833. Inspection of these two expressions for the economic cost criterion and some reflection tell us that there is *no* value of b for which our seaport manager is indifferent between the two inspection regimes being studied.

Figure 3 plots the economic cost criterion AWQ on the vertical axis against alternate values of b on the horizontal axis. Figure 3 tells us that AWQ is *always* lower when the less stringent L regime is used to inspect arriving ships in our seaport. In other words, for *all* values of b, which measures the difference in the variability—and hence reliability—of the two inspection regimes, there *is* a tension between economic cost minimization and inspection stringency or biological invasion damage control.

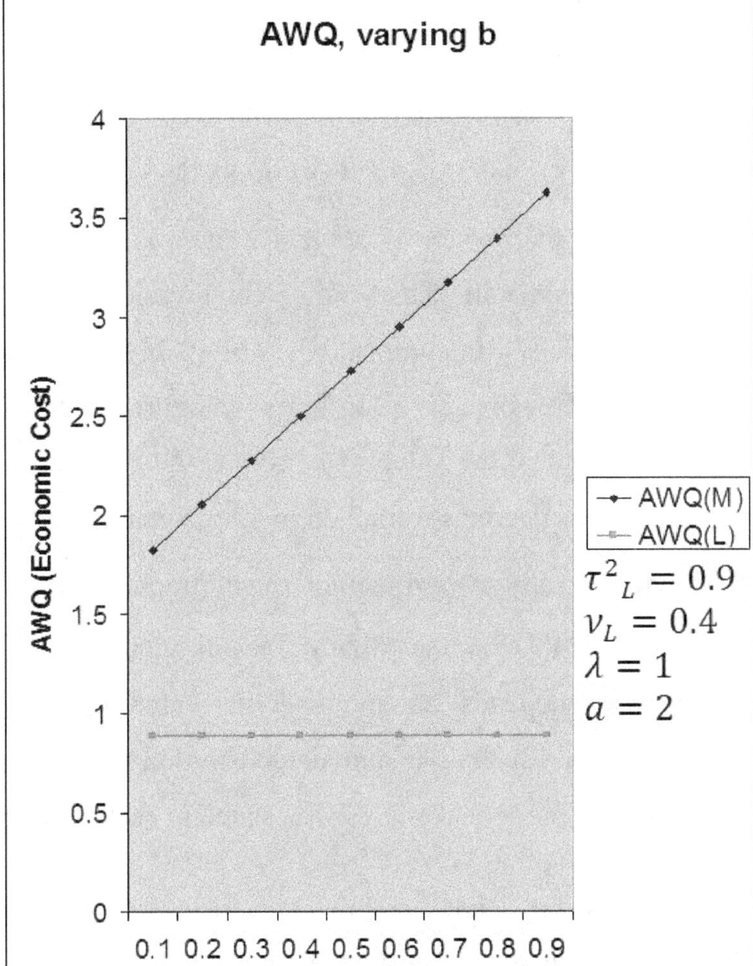

Figure 3

Does the above result depend on the specific value of a that we have chosen to conduct the numerical analysis with? The answer is yes, and, to see this, consider what would happen if, instead of setting a=2, we set a=1.2. In this case, once again substituting the values of the various parameters in equation (1), we get AWQ_M =0.2215+0.8654b and AWQ_L =0.8833. Now setting AWQ_M =0.2215+0.8654b=0.8833= AWQ_L, we get b=0.7647. This tells us that, when we set a=1.2 there is *no* tension between economic cost minimization and biological invasion damage control, as long as $b \in (0,0.7647]$. In contrast, when $b \in (0.7647,1)$, there *is* a tension between these two objectives. In terms of figure 3, what is happening is that, as we reduce the value of a from a=2 to a=1.2, the slope of the graph of AWQ_M declines from 2.25 to 0.8654 and the graph itself gets pulled down vertically toward the horizontal graph of AWQ_L. The reader should note that this finding—that the result obtained in the previous two paragraphs depends on the specific value of a—is consistent with the general illustration and demonstration of seaport specificity objectives of our numerical analysis, as stated in section 3.

Our analysis thus far leads to four conclusions. First, the theoretical analysis tells us that the question as to whether there is or isn't a tension between economic cost minimization and biological invasion damage control *cannot* be resolved unambiguously. Second, for many possible values of the a parameter and for all possible values of the b parameter when a=2, there *is* a tension between economic cost minimization and biological invasion damage control. Third, and at variance with the second conclusion, for several values of λ or the volume of maritime trade parameter, there is *no* tension between economic cost minimization and inspection stringency. Finally, by varying both a and b, we have graphically shown the impact that *alternate choices* of these two parameters have on the *AWQ* objective function. In addition, this variation exercise also shows that for these two parameters, inspection stringency, defined in terms of the mean and the variance of the inspection time random variable, lies in the *interval* $a \in (1, \infty)$ and $b \in (0,1)$. These four conclusions are consistent with the main findings obtained by DeAngelo *et al.* (2007) with the *AWS* cost criterion, and hence we note that the basic findings of DeAngelo *et al.* (2007) are general in the sense that they hold not only for the *AWS* cost criterion, but also for the *AWQ* cost criterion.

For real world applications of the model of this chapter, we would need to procure values for the arrival rate of ships and the averages and the variances of particular inspection regimes. In the United States, information about the arrival rate of ships can be procured from the administrative offices of individual seaports, such as Long Beach and, on occasion, from governmental agencies, such as the Office of Mobile Sources of the Environmental Protection Agency (EPA). Similarly, information about actual inspections in the United States can be procured from documents that are periodically produced by the

Congressional Research Service and from the Animal and Plant Health Inspection Service (APHIS).

4. Conclusions

Non-native animal and plant species often succeed in invading new habitats as a result of maritime trade in goods by means of ships. Hence, if an appropriate authority such as a seaport manager's goal is to prevent biological invasions, then he must inspect arriving ships for potentially deleterious biological organisms. Given this context, we used the *AWQ* cost criterion in the *M/G/*1 queuing model to investigate the generality of the findings obtained recently by DeAngelo *et al.* (2007). Our theoretical analysis showed that there is *no* unique answer to the question as to whether there is or isn't a tension between economic cost minimization and biological invasion damage control. In addition, our numerical analysis identified particular values of the essential model parameters for which there is a tension between economic cost minimization and biological invasion damage control. The general outcome of our combined theoretical and numerical analysis is twofold. First, the main findings of DeAngelo *et al.* (2007) are general in the sense that they hold for the *AWS* and for the *AWQ* cost criteria. Second, whether or not there is tension between economic cost minimization and biological invasion damage control depends greatly on the organizational details—the arrival rate of ships and the degree of stringency of inspections—in individual seaports.

The analysis in this chapter can be extended in a number of directions. We now suggest one such direction. One issue of interest concerns the analysis of situations in which, because of heavy traffic in a particular seaport, ships do not enter this seaport, but go instead to some other seaport. From the perspective of the manager of a single seaport, this heavy traffic situation can be analyzed with a queuing model in which "balking" is permitted. In other words, if an arriving ship finds n other ships already in our seaport, then this ship enters our seaport only with some probability p_n and it goes to some other seaport with probability $1-p_n$. If we assume that long queues discourage ships from entering a seaport, then we would expect p_n to be a decreasing function of n. A special case of this balking scenario is one in which a seaport has a finite capacity F so that $p_n=1$ for $n<F$ and $p_n=0$ for $n\geq F$. As in this chapter, one can analyze the properties of alternate inspection regimes in a model with these heavy traffic related features.

From the perspective of a social welfare maximizing manager, who is responsible for *all* seaports in a nation, a queuing analysis of the issues studied in this chapter would have to determine the desirable inspection stringency level at a particular seaport and *coordinate* inspection strategies in all the seaports in this nation. This way, differential inspection standards would not exist, and if they did exist, they would not, in and of themselves, provide incentives for arriving ships to favor certain seaports over others. If this coordination aspect of the problem is not addressed, then, in any particular nation, the risk of one or more biological invasions may actually increase, despite the presence of inspections in all national seaports. Research on maritime trade driven biological invasions that incorporate these aspects of the problem into the analysis will provide additional insights into a management problem that has considerable economic and ecological ramifications.

References

Arrow, K.J., and Chang, S.S.L. 1980. Optimal pricing, use, and exploration of uncertain natural resource stocks, in P.T. Liu, (Ed.), *Dynamic Optimization and Mathematical Economics*. Plenum Press, New York, NY.

Barbier, E.B. 2001. A note on the economics of biological invasions, *Ecological Economics*, 39, 197-202.

Batabyal, A.A. 2003. The persistence of ecological-economic systems: Alternate measures and their properties, *Annals of Regional Science*, 37, 323-336.

Batabyal, A.A. 2004. A research agenda for the study of the regulation of invasive species introduced unintentionally via maritime trade, *Journal of Economic Research*, 9, 191-216.

Batabyal, A.A., and Nijkamp, P. 2005. On container versus time based inspection policies in invasive species management, *Stochastic Environmental Research and Risk Assessment*, 19, 340-347.

Batabyal, A.A., Beladi, H., and Koo, W.W. 2005. Maritime trade, biological invasions, and the properties of alternate inspection regimes, *Stochastic Environmental Research and Risk Assessment*, 19, 184-190.

Batabyal, A.A., and Beladi, H. 2006. International trade and biological invasions: A queuing theoretic analysis of the prevention problem, *European Journal of Operational Research*, 170, 758-770.

Cox, G.W. 1993. *Conservation Ecology*. W.C. Brown Publishers, Dubuque, IA.

DeAngelo, G., Batabyal, A.A., and Kumar, S. 2007. On economic cost minimization versus biological invasion damage control, in A. Oude Lansink, (Ed.), *New Approaches to the Economics of Plant Health*. Springer, Heidelberg, Germany.

Eiswerth, M.E., and Johnson, W.S. 2002. Managing nonindigenous invasive species: Insights from dynamic analysis, *Environmental and Resource Economics*, 23, 319-342.

Eiswerth, M.E., and van Kooten, G.C. 2002. Uncertainty, economics, and the spread of an invasive plant species, *American Journal of Agricultural Economics*, 84, 1317-1322.

Feenstra, R.C. 2004. *Advanced International Trade*. Princeton University Press, Princeton, NJ.

Krugman, P.R., and Obstfeld, M. 2005. *International Economics*, 7th edition. Pearson Addison Wesley, Boston, MA.

Mangel, M. 1985. *Decision and Control in Uncertain Resource Systems*. Academic Press, Orlando, FL.

McAusland, C., and Costello, C. 2004. Avoiding invasives: Trade-related policies for controlling unintentional exotic species introductions, *Journal of Environmental Economics and Management*, 48, 954-977.

Office of Technology Assessment (OTA). 1993. *Harmful Non-Indigenous Species in the United States*. OTA-F-565, Washington, DC.

Olson, L.J., and Roy, S. 2002. The economics of controlling a stochastic biological invasion, *American Journal of Agricultural Economics*, 84, 1311-1316.

Parker, I.M., Simberloff, D., Lonsdale, W.M., Goodell, K., Wonham, M., Kareiva, P.M., Williamson, M.H., Von Holle, B., Moyle, P.B., Byers, J.E., and Goldwasser, L. 1999. Impact: Toward a framework for understanding the ecological effects of invaders, *Biological Invasions*, 1, 3-19.

Pimentel, D., Lach, L., Zuniga, R., and Morrison, D. 2000. Environmental and economic costs of nonindigenous species in the United States, *BioScience*, 50, 53-65.

Ross, S.M. 2003. *Introduction to Probability Models*, 8th edition. Academic Press, San Diego, CA.

Salvucci, S., Walter, E., Conley, V., Fink, S., and Saba, M. (1997). *Measurement Error Studies at the National Center for Education Statistics*. United States Department of Education, Washington, DC.

Taylor, H.M., and Karlin, S. 1998. *An Introduction to Stochastic Modeling*, 3rd edition. Academic Press, San Diego, CA.

Uhler, R.S., and Bradley, P.G. 1970. A stochastic model for determining the economic prospects of petroleum exploration over large regions, *Journal of the American Statistical Association*, 65, 623-630.

USDA, APHIS, and USFS. 2000. Pest Risk Assessment for Importation of Solid Wood Packing Materials into the United States. http://www.aphis.usda.gov/ppq/pra/swpm/complete.pdf

Vitousek, P.M., D'Antonio, C.M., Loope, L.L., and Westbrooks, R. 1996. Biological invasions as global environmental change, *American Scientist*, 84, 468-478.

Work, T.T., McCullough, D.G., Cavey, J.F., and Komsa, R. 2005. Arrival rate of nonindigenous insect species into the United States through foreign trade, *Biological Invasions*, 7, 323-332.

Yu, C.H. 2005. Reliability and validity. http://seamonkey.ed.asu.edu/~alex/teaching/assessment/reliability.html

Yang, Z., and Perakis, A.N. 2004. Multiattribute decision analysis of mandatory ballast water treatment measures in the US great lakes, *Transportation Research Part D*, 9, 81-86

CHAPTER 12

A THEORETICAL ANALYSIS OF RANDOM INSPECTIONS AND FINES IN INVASIVE SPECIES MANAGEMENT - WITH S.J. YOO

Abstract. In this chapter, we conduct a theoretical analysis of inspections in a stochastic environment and we shed light on two hitherto unstudied issues concerning inspections in the context of invasive species management. First, given a particular port of entry in a country, we study the properties of a *random* inspection scheme. Second, we compute the average total *fines* that will be collected in the long run by an inspection agency that uses the above inspection scheme to screen arriving ships for the presence of one or more invasive species.

1. Introduction

There is no gainsaying the fact that we now live in an era of globalization. The phenomena of globalization in general and shocking recent events involving terrorism in the United States (US), Spain, and the United Kingdom have generated great interest in issues concerning security across the world. In the US, in particular, this interest has manifested itself in the substantially increased interest in inspecting goods that are brought into the country from other parts of the world by means of airplanes, trucks, and, perhaps most notably, ships. This concern with seaport security, in particular, is not misplaced. As *The Economist* (Anonymous, 2006) has recently noted, only about five percent of the containers that bring two billion tonnes of cargo to US seaports are actually inspected. Therefore, it is not difficult at all for all kinds of illegal goods, and possibly detrimental animal and plant species, to get into the US.

Batabyal (2004), Work *et al.* (2005) and DeAngelo *et al.* (2007) have clearly demonstrated that, in addition to transporting goods between regions, airplanes, trucks, and ships have unwittingly also managed to carry all manner of invasive plant and animal species[63] from one part of the world to another. This inadvertent carriage has taken place in many different ways. Three examples follow. First, on occasion, invasive animal species have succeeded in lodging themselves in the landing gear of airplanes and, in this way, they have traveled as stowaways from one part of the world to another. Second, a number of marine invasive species have been introduced inadvertently into a particular part of the world by ships dumping their ballast water. Cargo ships often carry ballast water, in order to boost vessel stability when they are not carrying full loads. When these ships come into a seaport,

63 Invasive species are also referred to as alien species, as exotic species, and as non-native species.

this ballast water must be discarded before cargo can be loaded. Finally, and perhaps most significantly, ships and trucks have introduced invasive species into a particular part of the world by means of the containers they routinely use to carry cargo from one part of the world to another. In this context, the reader should understand that invasive species can remain concealed in containers for extended periods of time. In addition, the material, such as wood that is commonly used to pack the cargo in the containers, may itself contain invasive species.

Biological invasions of new habitats by non-native species have frequently resulted in great losses to society. For the US alone, the extent of these losses is massive. In this regard, Keller and Lodge (2007) have noted that the state of Indiana spends more than $600,000 each year to control a particular invasive species, namely, the *Eurasian watermilfoil.* Similarly, Kolar and Lodge (2001) have pointed out that the total costs of all invasive species is around $137 billion per year. In addition to these economic costs, invasive species have also given rise to serious biological damage. For instance, Vitousek *et al.* (1996) have demonstrated that invasive species can change ecosystem processes, act as vectors of diseases, and diminish biological diversity. Cox (1993) has pointed out that, out of 256 vertebrate extinctions with a known cause, 109 are the outcome of biological invasions. The implication of the discussion in this paragraph is clear. Invasive species have frequently been a great menace to society.

Broadly speaking, there are two kinds of managerial actions that one can take to control the spread of invasive species and their deleterious effects. These are *pre-invasion* and *post-invasion* actions. The purpose of pre-invasion actions is to preclude non-native species from invading a new habitat. In contrast, post-invasion actions are intended to control a non-native species, given that this species has already invaded a new habitat. In recent times, several researchers have analyzed a particular kind of pre-invasion action, namely, inspections. McAusland and Costello (2004) have shown that, when one considers the future effects of current invasive species introductions, one is led to a course of action that may involve the use of higher or lower tariffs but certainly involves more stringent inspections. Batabyal and Nijkamp (2005) have shown that, in an inspection cycle, the so called "container policy" is preferable to the so called "temporal policy," because the former policy leads to lower long run expected net costs from inspections. Using a model of seaport inspections, Batabyal (2006, 2008) has provided a rationale for and has developed aspects of the differential regulatory treatment of imports when invasive species are a potential problem. Batabyal and Yoo (2006) have analyzed the statistical properties of what they call a generic container inspection policy. Finally, DeAngelo *et al.* (2007) have used a queuing theoretic model of inspections to show that the question as to whether there is or is not a tension between the objectives of economic cost reduction and biological invasion damage control cannot be resolved unambiguously.

The papers discussed in the previous paragraph have surely advanced many aspects of our understanding of the role of inspections in invasive species management. Even so, there are two salient issues about inspections that have received *no* attention in the literature. Hence, in this chapter, we conduct a theoretical analysis of these two hitherto unstudied issues, in a stochastic environment. Specifically, in section 2.1, we describe a stylized model of inspections in which ships—possibly with injurious invasive species—arrive at a seaport in a country called Home. Next, in section 2.2, we study the properties of a *random* inspection scheme. Then, in section 2.3, we compute the average total *fines* that will be collected in the long run by an inspection agency that uses the above inspection scheme to screen arriving ships for the presence of invasive species. Finally, section 3 concludes and then makes suggestions for extending the research described in this chapter.

2. The Theoretical Model

2.1. Preliminaries

Consider a port of entry, such as a seaport in an arbitrary country called Home.[64] Our subsequent analysis is conducted from the perspective of an inspection agency that has been entrusted with the task

64 We stress that our subsequent analysis does not depend on the port of entry being a seaport. Our analysis would go through for land ports of entry—such as a border crossing or an airport—as well.

of inspecting arriving ships in this seaport for the presence of one or more invasive species. The reader should note that, as used in this chapter, the term "inspection" refers to the examination of the containers that are used by ships to transport cargo or to the examination of the ballast water occasionally held by arriving ships or to the examination of both containers and ballast water.

Home engages in goods trade with a whole host of nations and hence ships from these various nations arrive in our seaport in Home to unload and/or to load cargo. Now, as noted in Batabyal (2006, 2008), the risk of inadvertent biological invasions in Home typically varies by trading partner. Therefore, consistent with the analysis in Batabyal (2006, 2008), we suppose that our seaport inspection agency has distinct protocols for inspecting the arriving ships from distinct nations. Put differently, ships arriving from country *A* are treated differently than ships arriving from country *B,* for any two arbitrary countries *A* and *B*. Let us now delineate the random inspection scheme that is the first of two key issues that we are studying in this chapter.[65]

65This random inspection scheme is based on the "continuous sampling" plan first formulated by Dodge (1943) and subsequently extended by White (1966) and by Bebbington *et al.* (2003).

2.2. Random inspection scheme

We focus on the ships coming into the seaport under study from some arbitrary country—say country A—that is also a trading partner of Home. These ships come into the Home seaport over time and sequentially. We suppose that, on the basis of previously collected historical data, our seaport inspection agency has determined that there is a fixed probability p that a given arriving ship from country A will have one or more invasive species on it. Hence, such an arriving ship will fail to pass our agency's inspection. We also suppose that whether a particular ship from country A does or does not have a problem with invasive species does not depend on the status of any other ship arriving in the Home seaport from this same country A.

Our seaport inspection agency in Home proceeds as follows. Initially, it inspects every ship from country A until i consecutive ships are found not to have any invasive species on them. Once this happens, our agency then inspects only one out of every r ships from country A at *random* until another ship with one or more invasive species on it is discovered. When this happens, our agency reverts to one hundred percent inspections until i consecutive ships with no invasive species on them are found. The agency's inspection continues in this way. The task before us now is to compute the average fraction of all country A ships that are and are not inspected. In what follows, we shall use the acronyms *AFS* and *AFSN* to refer to these two averages. Our computation proceeds in three steps.

In the first step, let state D_k (k=0,1,2,...i-1) denote the k consecutive country *A* ships with no invasive species that have been found during the one hundred percent inspection part of the scheme. Also, let state D_i denote the fact that the inspection scheme under study is in the second stage in which one out of every r country *A* ships is being inspected randomly. Time n follows the *nth* ship, whether or not it is inspected. Now, the reader should note that the above described sequence of states is a Markov chain[66] with transition probability P_{jk} =*Prob*{*state is* D_k *after* n+1 *ships/ state is* D_j *after* n *ships*}. Mathematically, we have

$$P_{jk}=\begin{cases} p \; or \; k = 0, 0 \le j < i, \\ 1 - p \; for \; k = j + 1 \le i, \\ p/r for \; k = 0, j = i, \\ 1 - \left(\frac{p}{r}\right) for \; k = j = 1, \\ 0 \; otherwise \end{cases} \tag{1}$$

66For textbook accounts of Markov chains, the reader should consult Taylor and Karlin (1998, chapter 4) or Ross (2003, chapter 4).

In the second step, we specify the limiting probabilities for the Markov chain whose transition probabilities are given in equation (1) and then we solve the equations that are satisfied by these limiting probabilities. To this end, let π_k be the limiting probability that the stochastic system we are studying is in state D_k for k=0,1,2,…,i. To solve for these limiting probabilities, we have to specify the equations that these limiting probabilities satisfy. These equations are

$$p\pi_0 + p\pi_1 + \ldots + p\pi_{i-1} + (p/r)\pi_i = \pi_0, \tag{2}$$

$$(1\text{-}p)\pi_0 = \pi_1, \tag{3}$$

$$(1\text{-}p)\pi_1 = \pi_2, \tag{4}$$

and we keep going in this manner until we get to

$$(1\text{-}p)\pi_{i-1} + \{1\text{-}(p/r)\}\pi_i = \pi_i \tag{5}$$

and

$$\pi_0 + \pi_1 + \pi_2 + \ldots + \pi_i = 1. \tag{6}$$

Manipulating equations (3) through (5) we can tell that $\pi_k = (1-p)^k \pi_0$ for k=0,1,…,i-1. Similarly, simplifying equation (5), we get

$\pi_i=(r/p)(1-p)^i\pi_0$. Having ascertained π_k in terms of π_0 for k=0,1,…,i, we can now use equation (6) and then simplify the resulting expression to get

$$\pi_0=\frac{p}{1+(r-1)(1-p)^i},\ \pi_i=\frac{r(1-p)^i}{1+(r-1)(1-p)^i},$$

$$\pi_k=\frac{p(1-p)^k}{1+(r-1)(1-p)^i},\ k=0,1,\ldots i-1. \tag{7}$$

In the third step, we provide explicit closed-form expressions for the two averages of interest, that is, *AFS* and *AFSN* respectively. Because each ship is inspected when in states D_0, D_1,…, D_{i-1} but only one out of r ships is inspected in state D_i, we can infer that $AFSI=(\pi_0+\pi_1+\ldots+\pi_{i-1})+(1/r)\pi_i$. Simplifying this last expression and then using the fact that *AFSNI*=1-*AFSI,* we get

$$AFSI=\frac{1}{1+(r-1)(1-p)^i} \text{ and } AFSNI=\frac{(r-1)(1-p)^i}{1+(r-1)(1-p)^i}. \tag{8}$$

Equation (8) tells us that, in the random inspection scheme of this chapter, the average fraction of arriving ships from country *A* that are inspected depends fundamentally on the fixed probability *p* that a given arriving ship will have one or more invasive species on it and on the positive integer r describing the number of ships out of which one will be inspected at random in the second stage. It is straightforward to verify that when, either p=1 or r=1, our random inspection scheme becomes a deterministic scheme in which all arriving ships from country A are inspected by the agency. Finally, when p=0, the average fraction of ships that are inspected depends only on the positive integer r and as r increases (decreases), the average fraction of ships inspected decreases (increases). This completes the discussion of the first of two key issues that we are studying in this chapter. We now proceed to the second key issue. This involves computing the average total fines that will be collected in the long run by our inspection agency when it uses the random inspection scheme of this section to screen arriving ships for the presence of invasive species.

2.3. Average total fines

In the previous section, we described the way in which our random inspection scheme would work for ships arriving from a particular country A. However, it is clear that, in addition to country A, ships from many other countries—with which Home trades—also arrive in the seaport under study. Further, there is an inspection protocol in place for the ships from every relevant country. Having said this, the next question that arises concerns the status of ships that fail our agency's inspection. In practice, agencies responsible for the management of invasive species in many countries, such as Japan and the US, levy fines on non-compliant entities.[67] Therefore, in the remainder of this section, we suppose that ships—from all the pertinent countries—that fail our agency's inspection, are fined and that the magnitude of these fines depends on the extent to which a particular ship is not in compliance with existing laws and regulations in Home. Put differently, the magnitude (dollar value) of the individual ship fines are *random* variables.

To model this feature of the problem, we proceed as follows. At the beginning of each time period, ships from the various countries with which Home trades arrive in the seaport under study at the times of a renewal process with distribution law given by $F(x)$. We suppose that for every arriving ship, there is an inspector available to inspect this ship. Upon the completion of the inspection process, each ship pays a random fine to our agency and the amounts of this fine are described by the distribution law $G(y)$ where $y>0$.[68] Let $W(t)$ denote the total amount of all the fines from the various ships that have been collected by our

67For a more detailed corroboration of this claim, see Wanamaker (2008) and go to www.env.go.jp/en/nature/as/040427.pdf, www.state.hi.us/dlnr/Aliens3.html, and www.aphis.usda.gov/lpa/pubs/fsheet_faq_notice/fs_phcivilp.html.

68If we were to explicitly separate fine paying ships from non-fine paying ships, then the underlying mathematics would get unduly complicated. Therefore, to keep the subsequent mathematics straightforward, we are supposing that every ship pays a fine. The reader should not interpret this modeling feature literally. Put differently, the reader should interpret the "fines" paid by ships that pass inspection as a processing fee, and not as a punitive measure. Having said this, the salient point to note here is that the individual ship fines are random variables and we are explicitly modeling this point.

inspection agency by time t. The outstanding task before us now is to provide an explicit stochastic characterization of the total amount $W(t)$.

To this end, let Y_1, Y_2, Y_3,... denote the successive individual ship fines and let $N(t)$ denote the total number of ships that arrive in the Home seaport in the time interval $(0,t]$. We can now express the fine total $W(t)$ as a particular sum and that sum is

$$W(t)=\sum_{k=1}^{N(t)+1} Y_k . \tag{9}$$

The reader will note that the sum $W(t)$ in equation (9) is a random variable. Therefore, it makes sense to focus not on $W(t)$ *per se* but on its expectation $E[W(t)]$. Consistent with the discussion in section 1, the most convenient way to compute the above expectation would be to take a long run view of inspections and fines and compute the limiting expectation given by $\lim_{t\to\infty} E[W(t)]/t$. Now, the theory of renewal processes[69] tells us that $\lim_{t\to\infty} E[W(t)]/t=E[Y]/E[X]$, where Y and X denote the fines and time respectively. The two expectations on the right-hand-side (RHS) of the previous expression can be simplified

69For textbook expositions of renewal theory, the reader should consult Taylor and Karlin (1998, chapter 7) or Ross (2003, chapter 7).

further. This simplification gives us the limiting expectation we seek in its simplest form. Specifically, we get

$$\lim_{t\to\infty} \frac{E[W(t)]}{t} = \frac{E[Y]}{E[X]} = \frac{\int_0^\infty \{1-G(y)\}dy}{\int_0^\infty \{1-F(x)\}dx} \tag{10}$$

Equation (10) gives us a closed-form expression for the average total fines that will be collected in the long run by our inspection agency when it uses the random inspection scheme of section 2.2 to screen arriving ships for the presence of invasive species. The information contained in equation (10) can be used to facilitate the general task of invasive species management in two ways. First, this equation can be used to determine whether it is feasible to make the conduct of inspections by our agency a revenue-neutral operation. Put differently, the objective here would be to ascertain whether it is possible to meet the agency's costs with the revenue from the collected fines. Second, equation (10) can play the role of a constraint in an expected net social benefit from inspections maximization problem. The idea here would be to conduct ship inspections efficiently so that the net social benefit from inspections is maximized and, at the same time, the fine based revenue generated by these inspections does not fall below an exogenously given threshold. This concludes our discussion of the second key issue of this chapter.

3. Conclusions

In this chapter, we conducted a theoretical analysis of inspections in a stochastic environment and we shed light on two hitherto unstudied issues concerning inspections in the context of invasive species management. First, given a particular port of entry, we analyzed the properties of a random inspection scheme. Second, we computed the average total fines that will be collected in the long run by an inspection agency that uses the above inspection scheme to screen arriving ships for the presence of one or more invasive species.

The analysis in this chapter can be extended in a number of directions. Here are two suggestions for extending the research described in this chapter. First, we treated the probability p that an individual ship from a particular country will have one or more invasive species on it as exogenous to the analysis. Therefore, it would be useful to formally study the estimation of this important probability. Second, following the discussion towards the end of section 2.3, it would be useful to set up and solve an optimization problem involving the efficient allocation of inspection resources and the attainment of a threshold level of revenue from fines. Studies of inspections in invasive species management that incorporate these features of the problem into the analysis will provide further insights into a management function that has significant economic and ecological implications.

References

Anonymous. 2006. Trouble on the waterfront, *The Economist*, February 23.

Batabyal, A.A. 2004. A research agenda for the study of the regulation of invasive species introduced unintentionally via maritime trade, *Journal of Economic Research*, 9, 191-216.

Batabyal, A.A. 2006. A rationale for the differential regulatory treatment of imports when invasive species are a potential problem, *Studies in Regional Science*, 36, 179-187.

Batabyal, A.A. 2008. Two theoretical issues concerning the differential treatment of trading partners in alien species management, *Ecological Economics*, 64, 679-682.

Batabyal, A.A., and Nijkamp, P. 2005. On container versus time based inspection policies in invasive species management, *Stochastic Environmental Research and Risk Assessment*, 19, 340-347.

Batabyal, A.A., and Yoo, S.J. 2006. Some statistical properties of a generic container inspection policy in invasive species management, *Ecological Economics*, 60, 1-4.

Bebbington, M., Lai, C., and Govindaraju, K. 2003. Continuous sampling plans for Markov-dependent production processes under limited inspection capacity, *Mathematical and Computer Modelling*, 38, 1137-1145.

Cox, G.W. 1993. *Conservation Ecology*. W.C. Brown Publishers, Dubuque, Iowa.

DeAngelo, G.J., Batabyal, A.A., and Kumar, S. 2007. An analysis of economic cost minimization and biological invasion damage control, *Annals of Regional Science*, 41, 639-655.

Dodge, H.F. 1943. A sampling inspection plan for continuous production, *Annals of Mathematical Statistics*, 14, 264-279.

Keller, R.P., and Lodge, D.M. 2007. Species invasions from commerce in live aquatic organisms: Problems and possible solutions, *BioScience*, 57, 428-436.

Kolar, C.S., and Lodge, D.M. 2001. Progress in invasion biology: Predicting invaders, *Trends in Ecology and Evolution*, 16, 199-204.

McAusland, C., and Costello, C. 2004. Avoiding invasives: Trade-related policies for controlling unintentional exotic species introductions, *Journal of Environmental Economics and Management*, 48, 954-977.

Ross, S.M. 2003. *Introduction to Probability Models*, 8th edition. Academic Press, San Diego, California.

Taylor, H.M., and Karlin, S. 1998. *Introduction to Stochastic Modeling*, 3rd edition. Academic Press, San Diego, California.

Vitousek, P.M., D'Antonio, C.M., Loope, L.L., and Westbrooks, R. 1996. Biological invasions as global environmental change, *American Scientist*, 84, 468-478.

Wanamaker, T. 2008. New rules help stop invasive species, *Watertown Daily Times*, May 25.

White, L.S. 1966. The evaluation of H 106 continuous sampling plans under the assumption of worst conditions, *Journal of the American Statistical Association*, 61, 833-841.

Work, T.T., McCullough, D.G., Cavey, J.F., and Komsa, R. 2005. Arrival rate of nonindigenous insect species into the United States through foreign trade, *Biological Invasions*, 7, 323-332.

CHAPTER 13

ON THE UNLOADING OF CONTAINER CARGO IN A SEAPORT AND ITS INSPECTION FOR THE PRESENCE OF DELETERIOUS INVASIVE SPECIES

Abstract. We use a continuous-time Markov chain model to study the properties of a seaport container cargo unloading and inspection protocol. The purpose of cargo inspections is to detect, and thereby eliminate, the damage from stochastically arriving invasive species. Our analysis leads to five findings that can be used to, *inter alia*, optimize unloading and inspection activities. First, we determine the long run fraction of time that the unloading inspector is idle. Second, we compute the long run average number of ships from which cargo is unloaded per unit time. Third, we calculate the long run average number of ships that are waiting to have their cargo unloaded and inspected. Fourth, we ascertain the long run average waiting time per ship in our seaport. Finally, we use a simple numerical example to illustrate the computation of some of the above mentioned metrics.

1. Introduction

Invasive species (also known as alien and as non-native species) have been introduced into new habitats—either intentionally or unintentionally—for quite some time. Generally speaking, the intentional introductions have been less problematic and, today, most nations have in place an elaborate set of procedures that govern the circumstances under which invasive species may be introduced and the rules and regulations that must be followed by individuals seeking to undertake such intentional introductions.

In contrast, and as documented in many of the previous chapters in this book, the unintentional introductions have frequently proved to be extremely costly to the nations that have been at the receiving end of these introductions. Therefore, many nations now have in place an elaborate system of inspections at various border points and in interior points to reduce the known threat posed by non-native species that are successful in invading new habitats. In this regard, border points include airports, seaports, and land crossings, and interior points include specific locations where, as in Idaho in the United States, department of agriculture officials inspect watercraft to ensure that they are not infected with, *inter alia*, the zebra mussel.

Given the salience of *inspections* in controlling the spread of and the damage from invasive species, several researchers have now studied the utility of inspections as a management tool. McAusland and Costello (2004) analyze the role that inspections play in mitigating the damage caused by the unintentional introduction of invasive species stemming from international trade in commodities. They show that, when one takes a dynamic view and considers the future effects of current introductions, it is optimal to conduct stringent inspections. Batabyal (2006) uses queuing theory and analyzes a simple model of maritime trade and ship inspections in which there is a single importing nation called Home and two nations that export to Home. He constructs a rule which shows the circumstances under which it makes sense for a seaport manager in Home to grant favorable treatment to the exports of one or the other exporting nation. This rule depends on the average time it takes to inspect ships from the two exporting nations and on the dollar value of the imports from the two trading partners of Home. Moffitt *et al.* (2007) show how inspections for invasive species can be made robust when the underlying budget for inspections is limited. DeAngelo *et al.* (2007) study economic cost minimization and biological invasion damage control and generate two interesting conclusions. First, for many values of their model parameters that delineate the strictness of inspections, there is a tension between the twin objectives of economic cost reduction and inspection stringency. Second, and in contrast, for most values of their model parameter that depicts the volume of maritime trade handled by the seaport under study, there is no tension between economic cost minimization and inspection stringency.

From an invasive species management perspective, is it better to destroy or to treat contaminated shipments? This question is addressed by Ameden *et al.* (2007). Their analysis shows that, in response to increased inspections, importers may reduce the "due care" exercised by them. In addition, these researchers point out that the effects of enforcement depend significantly on the nature of the "due care" technology, as well as on the likelihood of detection. Merel and Carter (2008) study trade policies for managing the risk from invasive species that arrive with imports. They contend that the first best policy is a two-part tariff that involves levying a penalty on contaminated units which internalizes the invasive species externality and a fee on all imported units to cover inspection costs. Batabyal and Yoo (2008) study the properties of a random inspection scheme in a port of entry and then compute the mean total fines that will be collected in the long run by an inspection agency that uses the above mentioned inspection scheme to screen arriving ships for the presence of detrimental invasive species. Batabyal (2008) focuses on container inspections and derives the long run expected net cost of inspections (*LRENC*). He then contends that the optimal dependability and speed of container inspections is the solution to this *LRENC* minimization problem. Finally, Sanchirico *et al.* (2010) analyze a two region bioeconomic model of invasive species management with several control variables, such as inspections, removal, and sustainable land management. These researchers show that, *inter alia*, the non-linear interactions between the two regions are a salient determinant of which management regime is optimal.

The studies discussed in the preceding two paragraphs have certainly enhanced our comprehension of the many ways in which inspections can be a part of the optimal management of invasive species. Even so, it is fair to say that we still know relatively little about the management of invasive species in a *probabilistic* environment where the pertinent deleterious species originate in container cargo that is now a standard part of the transportation of goods between nations using ships. Given this lacuna in the extant literature, we use a continuous-time Markov chain model[70] to analyze the properties of a seaport container cargo unloading and inspection protocol. The purpose of cargo inspections is to detect and thereby eliminate the damage from *stochastically* arriving invasive species. Our examination leads to five conclusions that can be used to, *inter alia*, optimize unloading and inspection activities. First, we ascertain the long run fraction of time that the unloading inspector is idle. Second, we calculate the long run average number of ships from which cargo is unloaded per unit time. Third, we compute the long run mean number of ships that are waiting to have their cargo unloaded and inspected. Fourth, we determine the long

70See Ross (1996, pp. 231-294) or Tijms (2003, pp. 141-186) for textbook expositions of continuous-time Markov chains.

run mean waiting time per ship in our seaport. Finally, we use a simple numerical example to illustrate the computation of some of the above mentioned metrics.

The rest of this chapter is organized as follows. Section 2.1 delineates the continuous-time Markov chain model of a seaport container cargo unloading and inspection protocol. Section 2.2 first computes the long run fraction of *time* that the unloading inspector in the seaport is *idle,* and then it determines the long run average *number* of ships in our seaport from which container cargo *is* unloaded per unit time. Section 2.3 first demonstrates how we can determine the long run average *number* of ships that are *waiting* to have their cargo unloaded and inspected and then it derives the long run average waiting *time* per ship in our seaport. Section 2.4 provides a numerical example to show how some of the above mentioned metrics might be computed. Section 3 concludes and then discusses potential extensions of the research delineated in this chapter.

2. The Theoretical Framework

2.1. Preliminaries

Consider a seaport in a nation, such as the United States, that has a container terminal and, to keep the subsequent mathematical analysis straightforward, a single inspector. This inspector performs two basic functions. First, he is broadly in charge of unloading the container cargo that arrives in this seaport on ships from different parts of the world. Second, he is responsible for carefully inspecting the container cargo he has just unloaded for the presence of one or more deleterious invasive species. We suppose that our inspector can serve only one incoming ship at a time.

The time it takes to unload the container cargo from a single ship is assumed to be exponentially distributed with mean $1/\zeta_1$ where $\zeta_1 > 0$. After the container cargo from a particular ship has been unloaded, the ship in question leaves the seaport under study. The inspector then commences the key task of inspecting the unloaded container cargo. This inspection task takes a random amount of time to complete and we suppose that this random time is exponentially distributed with mean $1/\zeta_2$ where $\zeta_2 > 0$. After completing the inspection task, our inspector is available to unload the container cargo from the next ship that has arrived in our seaport. The time between the arrival of successive ships in our seaport is also random and is assumed to be exponentially distributed with mean $1/\theta$ with $\theta > 0$. For the purpose of this chapter, there are a finite number of $N > 0$ ships that visit our seaport.

To cast the description of events in the preceding two paragraphs in terms of a continuous-time Markov chain, let $X_1(t)$ denote the number of ships that are present before our unloading inspector at any time t. Let $X_2(t)$ be equal to one, if our inspector is in the process of inspecting the unloaded container cargo from a ship at any time t and zero otherwise. From the discussion in Ross (1996, pp. 231-294) or that in Tijms (2003, pp. 141-186), it is straightforward to confirm that the stochastic process $\{X_1(t),\ X_2(t)\}$ is a continuous-time Markov chain whose state space is given by I where $I=\{(i, j) \mid i=0,1,2,\ldots N; j=0,1\}$.

Now, if we denote the equilibrium probability of state (i, j) by $p(i, j)$, then the so called equilibrium or balance equations for the above described continuous-time Markov chain are given by[71]

$$N\theta p(0, 0)= \zeta_1 p(0, 1), \tag{1}$$

$$\{\zeta_1+(N-i)\theta\}p(i, 0)=(N-i+1)\theta p(i-1, 0)+ \zeta_2 p(i, 1),\ 1\leq i\leq N, \tag{2}$$

and

$$\{\zeta_2+(N-i)\theta\}p(i, i)=(N-i+1)\theta p(i-1, 1)+ \zeta_1 p(i+1, 0),\ 0\leq i\leq N, \tag{3}$$

with the stipulation that $p(N+1, 0)=p(-1, 1)=0$. Finally, we have the normalizing equation, which tells us that

71 See Ross (1996, pp. 252-254) or Tijms (2003, pp. 150-151) for additional details.

$$\sum_{i=0}^{i=N}\{p(i,0)+p(i,1\}=1. \tag{4}$$

Inspecting the balance equations (1)-(3), we see that these equations can be solved recursively. In particular, if we begin with $\bar{p}(0, 0)=1$, then we can successively calculate the probabilities $\bar{p}(0, 1)$, $\bar{p}(1, 0)$, $\bar{p}(1, 1)$ and finally the probabilities $\bar{p}(N\text{-}1, 0)$, $\bar{p}(N, 0)$, and $\bar{p}(N, 1)$. Having computed these probabilities, note that the probabilities $p(i, j)$ follow by normalizing the various $\bar{p}(i, j)'s$. This completes the delineation of our seaport container cargo unloading and inspection protocol. Our next tasks are to first calculate the long run fraction of *time* that the unloading inspector in the seaport is *idle* and then to ascertain the long run average *number* of ships in our seaport whose container cargo *is* unloaded per unit time.

2.2. Long run fraction of time and the long run average number of ships

Some thought tells us that the long run fraction of time that our container cargo unloading inspector is idle is given by the probability $p(0, 0)$. Now, observe that the long run average number of ships from which container cargo is unloaded per unit time is given by the long run average number of state transitions of our continuous-time Markov chain $\{X_1(t), X_2(t)\}$ per unit time from state $(i, 0)$ to state $(i\text{-}1, 1)$ for some $i\geq 1$.

Thus, if we define the average throughput in our model by T then we have[72]

$$T = \zeta_2 \sum_{i=1}^{i=N} p(i, 0). \tag{5}$$

A second way of characterizing the mean throughput (T) in our continuous-time Markov chain model is to recognize that the average number of ships whose container cargo is unloaded per unit time must be equal to the mean number of ships arriving in our seaport's unloading facility per unit time. Mathematically, this tells us that

$$T = \sum_{i=0}^{i=N} (N-i)\theta \{p(i, 0) + p(i, 1)\}. \tag{6}$$

The reader will note that the key uses of the idleness fraction $p(0, 0)$ and either equation (5) or (6) stem from the fact that they can, in principle, be used to reduce idleness in our seaport from the standpoint of the container cargo unloading inspector. Put differently, these two metrics can be used to increase the *efficiency* of the container cargo unloading and inspection protocol that we have been studying thus far. Our next two tasks are to first demonstrate how we can determine the

72Average throughput refers to the average long run number of ships that pass through our seaport for the purpose of first unloading their container cargo and then having this cargo be inspected for the presence of noxious invasive species. Therefore, the average throughput is a measure of our container cargo unloading and inspection protocol's performance.

long run average *number* of ships that are waiting to have their cargo unloaded and then to derive the long run average *waiting time* per ship in our seaport.

2.3. Number of waiting ships and the wait time per ship

The long run mean number of ships that are waiting to have their cargo unloaded can be computed by using a version of a well-known result from queuing theory known as Little's formula.[73] To see this, note that the long run average number of ships that are waiting to have their container cargo unloaded is the same as the long run average number of ships in queue in our seaport. Let us denote this long run number by L_q. Then, in the context of our problem, some thought tells us that

$$L_q = \sum_{i=1}^{i=N} (i-1)\{p(i, 0)+p(i, 1)\}. \tag{7}$$

Now, let W_q denote the long run average *waiting time* per ship in our seaport. The reader will note that this long run average can also be thought of as the long run average delay in queue per ship in the seaport

73 See Tijms (2003, pp. 50-51) for additional details on Little's formula.

under study. Also, let θ_T be the mean arrival rate of the container cargo carrying ships in our seaport. Mathematically, θ_T can be expressed as

$$\theta_T = \sum_{i=0}^{i=N} (N-i)\theta\{p(i,\ 0)+p(i,\ 1)\}. \tag{8}$$

With the above expression for θ_T in place, we can now apply equation 2.3.1 in Tijms (2003, p. 51) and conclude that the W_q metric we seek is

$$W_q = L_q / \theta_T, \tag{9}$$

where L_q is given by equation (7). Inspecting equations (7) and (9) we see that both the expressions in these two equations delineate alternate notions of waiting in our seaport. Therefore, by minimizing one or both of these two notions of waiting, it will once again be possible in principle to improve an attribute of the container cargo unloading and inspection protocol that is the subject of this chapter. We now proceed to the last section of this chapter. Our objective in this section is to provide a numerical example to show how some of the metrics of sections 2.2 and 2.3 might be computed.

2.4. A numerical example

Let ζ_1=1/3, ζ_2=2 θ=1/50, and N=10. Then, laborious but straightforward calculations using equation (7) tell us that the long run average number of ships that are waiting to have their container cargo unloaded in our seaport or L_q=0.590. Undertaking similar computations with equation (8) tells us that the mean arrival rate of the container cargo carrying ships in our seaport or θ_T=0.178. Finally, using the above two values for L_q, θ_T, and equation (9), we get an expression for the long run average waiting time per ship in our seaport. That expression is W_q =0.590/0.178=3.315 time units.

This completes our numerical example. More generally, the above example shows that, with the requisite data for the three model parameters ζ_1, ζ_2, θ, and the number of ships N, the various container cargo unloading and inspection protocol performance metrics delineated in this chapter can be computed and then operationalized.

3. Conclusions

In this chapter, we used a continuous-time Markov chain model to analyze the attributes of a seaport container cargo unloading and inspection protocol. The objective of cargo inspections was to detect, and thereby eliminate, the damage from probabilistically arriving invasive species. Our analysis led to five findings that can be used to optimize unloading and inspection activities. First, we determined the long run proportion of time that the unloading inspector is idle. Second, we calculated the long run average number of ships from which cargo is unloaded per unit time. Third, we ascertained the long run average number of ships that have been waiting to have their cargo unloaded and inspected. Fourth, we computed the long run average waiting time per ship in our seaport. Finally, we used a simple numerical example to illustrate how, given data, the various performance metrics associated with the cargo unloading and inspection protocol of this chapter, might be operationalized.

The analysis described in this chapter can be extended in a number of different directions. Here are two examples of such extensions. First, consistent with the discussion towards the end of section 2.2, it would be useful to determine what concrete actions might be taken to alter the probabilities described in equations (5) and (6) and thereby improve the *efficiency* of the container cargo unloading and inspection protocol that we have studied in this chapter. Second, and consistent with the discussion in section 2.3, it would be interesting to first assign explicit benefit and cost terms to the expressions in equations (7) and (9) and then to select apposite arguments to *minimize* the resulting benefit and cost weighted waiting expressions. Studies of container cargo unloading and inspection protocols that analyze these aspects of the problem are sure to shed light on the desirability of some key prophylactic practices that are frequently vital in the efficient and optimal management of invasive species.

References

Ameden, H.A., Cash, S.B., and Zilberman, D. 2007. Border enforcement and firm response in the management of invasive species, *Journal of Agricultural and Applied Economics*, 39, 35-46.

Batabyal, A.A. 2006. A rationale for the differential regulatory treatment of imports when invasive species are a potential problem, *Studies in Regional Science*, 36, 179-187.

Batabyal, A.A. 2008. A theoretical note on the optimal dependability and speed of container inspections in invasive species management, *Studies in Regional Science*, 38, 199-204.

Batabyal, A.A., and Yoo, S.J. 2008. A theoretical analysis of random inspections and fines in invasive species management, *Economics Bulletin*, 17, 1-9.

DeAngelo, G.J., Batabyal, A.A., and Kumar, S. 2007. An analysis of economic cost minimization and biological invasion using the AWQ criterion, *Annals of Regional Science*, 41, 639-655.

McAusland, C., and Costello, C. 2004. Avoiding invasives: Trade-related policies for controlling unintentional exotic species introductions, *Journal of Environmental Economics and Management*, 48, 954-977.

Merel, P.R., and Carter, C.A. 2008. A second look at managing import risk from invasive species, *Journal of Environmental Economics and Management*, 56, 286-290.

Moffitt, L.J., Stranlund, J.K., Field, B.C., and Osteen, C.D. 2007. Robust inspection for invasive species with a limited budget, in A.G.J.M. Oude Lansink, (Ed.), *New Approaches to the Economics of Plant Health*. Springer, Dordrecht, The Netherlands.

Ross, S.M. 1996. *Stochastic Processes*, 2nd edition. Wiley, New York, NY.

Sanchirico, J.N., Albers, H.J., Fischer, C., and Coleman, C. 2010. Spatial management of invasive species: Pathways and policy options, *Environmental and Resource Economics*, 45, 517-535.

Tijms, H.C. 2003. *A First Course in Stochastic Models*. Wiley, Chichester, UK.

CHAPTER 14

OPTIMAL RESOURCE MANAGEMENT IN THE PRESENCE OF A DELETERIOUS ALIEN SPECIES: A STOCHASTIC MODEL FOR AN ORCHARD - WITH P. NIJKAMP

Abstract. Resources, such as orchards, are vulnerable to the detrimental effects of successful invasions by alien animal or plant species. To outline an appropriate policy response, we first use renewal theory to construct a stochastic model of optimal orchard management in the presence of a deleterious alien species. Next, we derive the orchard manager's long run expected cost (*LREC*) of orchard management per unit time. Finally, we show that when confronted with a successful biological invasion, the optimal number of trees that need to be removed and replanted in order to keep the orchard under study sustainable in the long run minimizes the *LREC* function mentioned above.

1. Introduction

In contemporary times, all over the world, various kinds of fruits and nuts, such as apples, oranges, peaches, and pecans, are grown in orchards. As a result, orchards are now key generators of agricultural revenue for their owners and/or managers in disparate parts of the world. Given the work of Batabyal (1999), Batabyal and Yoo (2007a, 2007b), and others, researchers now recognize that the present and the future state of an orchard is defined by factors that are partly economic and partially ecological in nature. The economic factors are rather straightforward. Specifically, orchards are the outcome of human actions that have been taken to promote the commercial and the non-commercial consumption of fruits and nuts. The ecological factors influencing orchards include weeds, inclement weather, and, most notably for our purpose, alien species. Using the parlance of modern researchers, orchards are *jointly determined* ecological-economic systems.[74]

The extant literature on orchards has analyzed many aspects of their management. For instance, Khushk and Smith (1996) have investigated the consequences of mango orchardists in Sindh, Pakistan, selling their harvesting rights to contractors. Davis *et al.* (1997) have shown that the unevenness in yields that is commonly encountered in peach production in the United States can be reduced by expanding the distances between peach orchards. Park and Florkowski (2003) have analyzed the impact that orchard conditions and other factors have on the decision to utilize new varieties in peach production in the United States. Darroch *et al.* (2004) have studied the effect of alternate investment decisions on the profitability of apple orchard owners in

74Following the work of Perrings (1987), ecological-economic systems are sometimes also known as economy-environment systems.

South Africa. Florkowski and Sarmiento (2005) have analyzed the nexus between the prices received by pecan growers and the attributes of their orchards in the United States. Seavert *et al.* (2006) have shown how crop profitability analysis (CPA) can be used to analyze the potential profitability of, *inter alia*, orchard crops. Forster and Gilligan (2007) have studied the control of disease infestations at the landscape scale and have shown that the optimal long term strategy requires an intermediate level of control that leads to the near eradication of the pertinent disease. Finally, Rahim *et al.* (2007) have used what they call a "real options approach" to investigate a farmer's incentives for abandoning or expanding gum production. On the basis of their analysis, these researchers conclude that deforestation and an expansion of gum forests and/or agroforests in the near future are not expected.

The papers that we have just commented on in the preceding paragraph have certainly increased our comprehension of several aspects of orchard management. Even so, it is important to recognize the following two points. First, orchards are not only jointly determined ecological-economic systems but their growth and output are also *dynamic* and *stochastic* in nature. Therefore, sustainable orchard management is fundamentally all about the management of uncertainty over time. Second, as observed by Acquaye *et al.* (2005) and Pierre *et al.* (2006), orchards are often invaded by one or more animal or plant alien species.[75] Hence, effective orchard management ought to take this aspect of the management problem into consideration, explicitly. Unfortunately, the papers discussed in the preceding paragraph and indeed the extant literature in general have *not* satisfactorily accounted for these two points.

Given this state of affairs, our central objective in this chapter is to use renewal theory[76] to construct and analyze a dynamic and stochastic model of optimal orchard management that explicitly takes into account the damage from a successful invasion by an alien animal or plant species.[77] To this end, we first derive an orchard manager's long run expected cost *(LREC)* of orchard management per unit time. Next, we focus on the optimal number of trees that need to be removed and replanted to keep the orchard under study sustainable in the long run and we show that this optimal number is the solution to the above mentioned *LREC* minimization problem. Now, before we move to the organizational details of the chapter, let us understand two key points.

75Citrus canker and the Mediterranean fruit fly are two examples of alien species that frequently have a negative impact on fruit orchards in Florida, in the United States. See Acquaye *et al.* (2005) and Pierre *et al.* (2006) for more information on this topic. In addition, the reader should note that alien species represent one kind of potential damage that orchard managers have to occasionally contend with. Other sources of potential damage include inclement weather, high winds, and native pests.

76See Ross (2003, pp. 401-473) and Tijms (2003, pp. 33-79) for textbook treatments of renewal theory.

77The reader should note that the orchard management problem can also be thought of as a land management problem, because an orchard represents a particular kind of land use. In addition, broadly speaking, there are two strategies for dealing with invasive species. The first strategy involves the social management of the risk of introductions. The second strategy involves defensive adaptive responses to introductions that are taken privately by an individual, such as an orchard manager. This chapter is about the use of this second kind of strategy.

First, although we are casting our dynamic and stochastic model as a model of orchard management, our modeling approach is actually quite general. In particular, this approach can usefully be thought of as a way of analyzing the "repair" of environmental damage in systems that can be described by a "renewal process" (on which more below). Therefore, our modeling approach is not specific to orchards and nor is it specific to the management of alien species. Indeed, in addition to crops that are typically grown in orchards, this approach can be applied to study the management of, for instance, berry and vineyard crops. Second, it should be noted that whether or not the solution referred to in the previous paragraph is sustainable depends not only on what the orchard manager of this chapter is doing but also on what other orchard managers-operating in the vicinity of the orchard under study in this chapter-are doing. Put differently, the solution proposed in this chapter is sustainable in a private sense, and *not* in a social sense.

The rest of this chapter is organized as follows. First, section 2.1 describes the renewal-reward theorem that will form the centerpiece of our orchard manager's *LREC* determination problem. Second, section 2.2 derives the *LREC* of orchard management per unit time. Third, section 2.3 solves our orchard manager's management problem, i.e., the *LREC* minimization problem and shows that the optimal number of trees that need to be removed and replanted to keep the orchard under study sustainable in the long run is the solution to this *LREC* minimization problem. Finally, section 3 concludes and discusses potential extensions of the research delineated in this chapter.

2. The Theoretical Framework

2.1. Preliminaries

The textbook by Ross (2003, pp. 416-425) tells us that a stochastic process $\{Z(t):t\geq 0\}$ is a counting process if $Z(t)$ represents the total number of counts that have taken place by time t. Clearly, since $Z(t\text{-}1)$, $Z(t)$, $Z(t+1)$, etc. are stochastic, the time between any two counts $Z(t)$ and $Z(t\text{-}1)$ is also stochastic. This time between any two counts is called the interarrival time. A counting process for which the interarrival times have a *general* cumulative probability distribution function is a renewal process.[78]

Consider a renewal process $\{Z(t):t\geq 0\}$ with interarrival times X_z $z\geq 1$, which have a cumulative probability distribution function $G(\bullet)$. In addition, assume that a monetary reward R_z is earned when the zth renewal is completed. Let $R(t)$, the total reward earned by time t, be

78In the context of the management of invasive species, there is typically not much basis for an individual, such as an orchard manager, to choose a particular cumulative probability distribution function for the interarrival times. This is why we are working with renewal processes for which the cumulative probability distribution function is general.

$\sum_{z=1}^{\angle(t)} R_z$, and let $E[R_z]=E[R]$, and $E[X_z]=E[X]$. The renewal-reward theorem-see Ross (2003, p. 417) or Tijms (2003, p. 41)-tells us that if $E[R]$ and $E[X]$ are finite, then with probability one,

$$\lim_{t\to\infty} \frac{E\left[R(t)\right]}{t} = \frac{E\left[R\right]}{E\left[X\right]}. \tag{1}$$

In words, equation (1) is telling us that if we think of a cycle being completed every time a renewal occurs, then the long run expected reward-the left-hand-side (LHS) of equation (1)-is the expected reward in a cycle or *E*[*reward per cycle*] divided by the expected amount of time it takes to complete that cycle or *E*[*length of cycle*]. The reader should note that the renewal-reward theorem describes a *long run* or *steady state* result and this theorem holds for positive rewards such as profits and for negative rewards such as costs. Let us now derive the long run expected cost per unit time or *LREC* for our orchard manager.

2.2. LREC of orchard management

Consider an orchard that consists of N independent, and, for all practical purposes, identical, trees that is overseen by a single individual decision maker called the orchard manager. The individual trees in this orchard grow independently of each other. An alien species, such as citrus canker or the Mediterranean fruit fly may invade this orchard and if such an invasion does, in fact, occur then we suppose that the trees in our orchard will get infected with a disease at a constant infection rate-over the useful life of the orchard under study-given by $\mu>0$. Now, consistent with actual practice,[79] our orchard manager will actually remove the infected trees from the orchard and plant new trees in their place only when the number of infected trees in this orchard reaches the critical number C, where $1 \leq C \leq N$.

To reiterate, if the number of infected and hence diseased trees reaches the critical number C, then our manager simultaneously removes the C infected trees from the orchard and plants new trees in their place. In what follows, to keep the mathematical analysis tractable, we suppose that this simultaneous tree removal and replanting exercise takes a negligible amount of time. Clearly, this tree removal and replanting exercise is costly. As such, we suppose that the total cost of the above exercise is part fixed, and part variable, in nature. Denote the fixed cost component by $F>0$ and the variable cost component by αC, where $\alpha>0$ is the marginal cost of removing and replanting trees. Therefore, in symbols, the total cost of removing and replanting C trees simultaneously is $F+\alpha C$. In addition to the fixed and the variable costs

79For more on these matters, go to http://edis.ifas.ufl.edu/CG040 and see, for instance, the 2007 Florida Citrus Pest Management Guide: *Citrus Canker*.

that we have just delineated, we suppose that there is an additional "diseased tree" cost of $\beta > 0$ per unit time for each infected and diseased tree in the orchard, before it is removed by our manager.

Given the above description of the orchard and our orchard manager's prophylactic activities, the outstanding task before us now is to use the renewal-reward theorem (equation (1)) to derive the *LREC* of orchard management. However, before we can use the renewal-reward theorem, we must first specify the apposite regenerative process that is the object of our study. In this regard, the reader should note that the continuous time stochastic process describing the number of uninfected or healthy trees is the pertinent regenerative process. In addition, the regeneration epochs in our model are those time points at which our manager removes the *C* infected trees from the orchard and plants new trees in their place.[80]

We begin by computing the expected length of the relevant renewal cycle or *E*[*length of cycle*]. In this regard, the reader will note that a random variable that is defined to be the minimum of two exponentially distributed random variables with means, say, $1/\delta_1$ and 1/

80The reader should pay careful attention to our definition of the relevant regenerative process and to the regeneration epochs. This specification and the long run or steady state nature of the analysis together explain the form of the expected cost per cycle in equations (3) and (4) below. In addition, note that a goal of this chapter is methodological. We want to show that renewal theory in general and the renewal-reward theorem in particular can be used to derive a meaningful objective function that an orchard manager might optimize. Given this goal, the only reasonable way to define both the regeneration process and the regeneration epochs is as we have defined them in this paragraph.

δ_2 is itself exponentially distributed with mean $1/(\delta_1+\delta_2)$.[81] In addition, we know that an exponentially distributed random variable has the memoryless property.[82] Using these two pieces of information together, we deduce that

$$E[\textit{length of cycle}]=\frac{1}{N\mu}+\frac{1}{(N-1)\mu}+\ldots$$

$$+\frac{1}{(N-C+1)\mu}=\sum_{i=0}^{i=C-1}\frac{1}{(N-i)\mu}. \tag{2}$$

In our model, the expected length of a cycle increases with the number of trees that might be infected. This result arises because the individual trees grow independently of each other and because we have used a constant infection rate given by $\mu>0$. This constant infection rate assumption is not atypical and it is important to note that such an assumption has been used previously to study infections in a variety of settings by Gough (1977), Roy (2000), and Onipchenko (2004).

81 See Taylor and Karlin (1998, pp. 46-47) for more on this basic result.

82 See Ross (2003, pp. 272-275) and Tijms (2003, pp. 440-441) for more on the memoryless property of exponentially distributed random variables.

Our next task is to compute the expected cost per renewal cycle or $E[cost\ per\ cycle]$. Our previous definition of the regeneration epochs in our model and some thought together tell us that the expectation we seek is given by

$$E[cost\ per\ cycle] = \frac{\beta}{(N-1)\mu} + \frac{2\beta}{(N-2)\mu} + \ldots$$

$$+ \frac{(C-1)\beta}{(N-C+1)\mu} = \sum_{i=0}^{i=C-1} \frac{i\beta}{(N-i)\mu}. \tag{3}$$

Having computed the expected cost to the orchard manager in a cycle and the expected length of this cycle, a straightforward application of the renewal-reward theorem-equation (1)- gives us our *LREC*. Specifically, we get

$$LREC = \frac{E[cost\ per\ cycle]}{E[length\ of\ cycle]} = \frac{\sum_{i=0}^{i=C-1} \frac{i\beta}{(N-i)\mu}}{\sum_{i=0}^{i=C-1} \frac{1}{(N-i)\mu}}. \tag{4}$$

Inspection of equation (4) tells us that our orchard manager's *LREC* depends fundamentally on the number of trees in the orchard (*N*), the critical number of trees that are infected before prophylactic actions commence (*C*), the alien species driven tree infection rate (μ), and finally the cost resulting from the presence of diseased trees in the orchard in a renewal cycle before our manager acts to remove them (β).[83]

Having derived the *LREC*,, let us now clearly point out that the purpose of this chapter is *not* to engage in a comparative analysis of the management of alien and native pests. Instead, and as noted previously in section 1, our objective is to show how renewal theory can be used to construct and analyze a dynamic and stochastic model of optimal orchard management that explicitly takes into account the damage from a successful invasion by an alien animal or plant species. Having said this, we now briefly comment on the use of this chapter's renewal theoretic framework to study the management of invasive *and* native pests.

The reader should note that the scope of renewal theory in general is broad, and hence this theory can certainly be used to study the management of both alien as well as native pests, in the context of either

83The reader will note that the expression for the *LREC* in equation (4) does not contain either the fixed cost term $F>0$ or the marginal cost term $\alpha>0$. This is not because we are ignoring these two terms. Instead, this is because of two factors. The first factor is the way in which we have defined the pertinent regeneration process that we are studying and the associated regeneration epochs (also see footnote 80). The second factor is that the *LREC* in equation (4) is based on the renewal-reward theorem and this theorem is a long run or steady state result. These two factors together tell us that in the steady state, the only cyclical cost term that matters is the $\beta>0$ term. This explains the absence of the $F>0$ and the $\alpha>0$ terms in equation (4). In addition, the reader should also note that our analysis in this chapter is stochastic but it is occurring in the steady state. Hence, we are not interested in analyzing the optimal timing of the replacement of diseased trees or the related irreversibility effect. For an eminently readable general discussion of these last two issues, see Dixit and Pindyck (1994). For a discussion of these last two issues in a natural resource setting, see Rahim *et al.* (2007).

an orchard or some other resource. In addition, because we use the renewal-reward theorem in particular in this chapter, it is important to recognize that this theorem can be used to derive meaningful objective functions that are relevant for the management of both alien and native pests. The differences between a renewal-reward theorem driven model of alien pest management and a renewal-reward theorem driven model of native pest management would lie in the details. As far as these details are concerned, three specific differences come to mind. First, the infection rates-μ in our model-for alien and native pests will typically be different. In particular, even if the rates are constant, we would expect to see unequal rates. Second, the fixed cost of removing and replanting trees-F in our model-will be different. Third, the marginal cost of removing and replanting trees-α in our model-will be dissimilar. Because of these three expected differences, even if we define the underlying regenerative process similarly, the expected cost per cycle and the expected length of a cycle for alien as opposed to native pests will typically be different. We now use equation (4) and discuss our orchard manager's optimization problem.

2.3. The optimization problem

Our orchard manager's objective is to choose the critical number of infected and diseased trees (C) that necessitate his prophylactic tree removal and replanting actions.[84] Mathematically, the optimization problem of interest is[85]

$$\min_{\{C\}}\left[\frac{\sum_{i=0}^{i=C-1}\frac{i\beta}{(N-i)\mu}}{\sum_{i=0}^{i=C-1}\frac{1}{(N-i)\mu}}\right]. \tag{5}$$

After several algebraic steps, the first-order necessary condition for an optimum to this problem can be written as

$$[\sum_{i=0}^{i=C-1}\frac{1}{(N-i)\mu}](\frac{\beta}{\mu})\frac{d}{dC}\{\frac{C-1}{N-C+1}\}=$$

$$[\sum_{i=0}^{i=C-1}\frac{i\beta}{(N-i)\mu}](\frac{1}{\mu})\frac{d}{dC}\{\frac{1}{N-C+1}\}. \tag{6}$$

84As noted in the last two paragraphs of section 2.2, the expected cost per cycle for the alien and the native pest management problems will typically be different because of dissimilar fixed and marginal costs terms. Hence, the resulting *LREC* expressions will be different. As such, the optimal values of the critical number of trees that are removed and replanted in these two management problems will also typically be different. This can be shown definitively if we use specific numerical values for all the germane model parameters. However, as we have already noted in the last two paragraphs of section 2.2, our goal in this chapter is not to engage in a comparative analysis of the management of alien and native pests.

85Because the orchard manager's choice variable C is a positive integer, the optimization problem given in equation (5) is, strictly speaking, an integer programming problem. Even so, following the approach employed in Batabyal and Beladi (2006), we use calculus to characterize the optimal solution to this problem. The reader will note that in following this approach, we are implicitly assuming that the optimal integer C can be closely approximated by the optimal continuous C.

Equation (6) tells us that the optimal critical number of trees C^* - that must be infected and diseased before our manager takes his prophylactic actions-ought to be chosen in a particular way. Specifically, our manager's optimal choice equalizes the marginal impact of this choice on the weighted expected length of the relevant renewal cycle (the LHS) with the corresponding marginal impact on the weighted expected management cost in this same renewal cycle (the RHS). In order to make additional progress with the optimality condition given in equation (6), one will typically have to make use of numerical methods. This concludes our discussion of the orchard manager's *LREC* minimization problem.

3. Conclusions

In this chapter, we used renewal theory to shed light on a simple stochastic model of optimal orchard management in the presence of a detrimental alien species. Specifically, from the standpoint of an orchard manager, we derived the *LREC* of management function. We then showed that the critical number of trees in the orchard that must be infected and diseased before the manager takes his prophylactic tree removal and replanting actions is the solution to a particular minimization problem in which the objective function is the above derived *LREC* function.

The analysis in this chapter can be extended in a number of directions. Here are two suggestions for extending this chapter's research. First, following the discussion in section 2.3, it would be useful to conduct numerical analyses to study the nature of the dependence of the *LREC* function on the choice of the optimal critical number of trees or C^*. Second, it would also be useful to eschew the use of the exponential distribution and analyze a model in which the biological invasion driven tree infection and disease rate is variable and not constant. Studies of the optimal management of orchards that incorporate these features of the problem into the analysis will provide additional insights into a management function that has salient economic and ecological ramifications.

References

Acquaye, A.K.A., Alston, J.A., Lee, H., and Sumner, D.A. 2005. Economic consequences of invasive species policies in the presence of commodity programs: Theory and application to citrus canker, *Review of Agricultural Economics*, 27, 498-504.

Batabyal, A.A. 1999. Aspects of the optimal management of cyclical ecological-economic systems, *Ecological Economics*, 30, 285-292.

Batabyal, A.A., and Beladi, H. 2006. International trade and biological invasions: A queuing theoretic analysis of the prevention problem, *European Journal of Operational Research*, 170, 758-770.

Batabyal, A.A., and Yoo, S.J. 2007a. A probabilistic approach to optimal orchard management, *Ecological Economics*, 60, 483-486.

Batabyal, A.A., and Yoo, S.J. 2007b. A stochastic analysis of the Holling resilience of an orchard, *Ecological Economics*, 64, 1-4.

Darroch, M.A.G., Hardman, P.A., and Ortmann, G.F. 2004. An assessment of apple orchard investments in South Africa under uncertainty and irreversibility, *South African Journal of Economic and Management Sciences*, 7, 601-615.

Davis, S.B., Price, J.T., Wetzstein, M.F., and Rieger, M.W. 1997. Reducing yield variation in peach orchards by geographic scattering, *American Journal of Agricultural Economics*, 79, 1119-1126.

Dixit, A.K., and Pindyck, R.S. 1994. *Investment Under Uncertainty*. Princeton University Press, Princeton, New Jersey.

Florkowski, W.J., and Sarmiento, C. 2005. The examination of pecan price differences using spatial correlation estimation, *Applied Economics*, 37, 271-278.

Foster, G.A., and Gilligan, C.A. 2007. Optimizing the control of disease infestations at the landscape scale, *Proceedings of the National Academy of Sciences*, 104, 4984-4989.

Gough, K.J. 1977. The estimation of latent and infection periods, *Biometrika*, 64, 559-565.

Khushk, A.M., and Smith, L.E.D. 1996. A preliminary analysis of the marketing of mango in Sindh province, Pakistan, *Pakistan Development Review*, 35, 241-255.

Onipchenko, V.G. Ed. 2004. *Alpine Ecosystems in the Northwest Caucasus*. Springer, Heidelberg, Germany.

Park, T.A., and Florkowski, W.J. 2003. Selection of peach varieties and the role of quality attributes, *Journal of Agricultural and Resource Economics*, 28, 138-151.

Perrings, C. 1987. *Economy and Environment*. Cambridge University Press, Cambridge, UK.

Pierre, R., Spreen, T.H., and Moss, C.B. 2006. Invasive species and biosecurity cost: Cost of monitoring and controlling Mediterranean fruit flies in Florida, *Journal of Agricultural and Applied Economics*, 38, 337-343.

Rahim, A.H., van Ierland, E.C., and Wesseler, J. 2007. Economic incentives for abandoning or expanding gum Arabic production in Sudan, *Forest Policy and Economics*, 10, 36-47.

Ross, S.M. 2003. *Introduction to Probability Models*, 8th edition. Academic Press, San Diego, California.

Roy, D. 2000. *Plant Breeding*. Alpha Science International, Oxford, UK.

Seavert, C., Hinman, H., and Klonsky, K. 2006. A crop profitability analysis for long-term crop investments, *Agricultural Finance Review*, 66, 125-135.

Taylor, H.M., and Karlin, S. 1998. *An Introduction to Stochastic Modeling*, 3rd edition. Academic Press, San Diego, California

Tijms, H.C. 2003. *A First Course in Stochastic Models.* Wiley, Chichester, UK.

CHAPTER 15

AN APPROACH TO THE MANAGEMENT OF ORCHARDS THAT ARE VULNERABLE TO ATTACKS BY INVASIVE SPECIES - WITH C.D. LOMINAC

Abstract. Orchards are vulnerable to the damaging effects of attacks by invasive species. Therefore, in this chapter, we focus on a representative tree in a stylized orchard and show how the theory of discrete time Markov chains can be used to construct a model of this tree that is to be managed in the possible presence of one or more injurious invasive species. Specifically, we accomplish three orchard management related tasks. First, we specify an apposite discrete time Markov chain theoretic model for our representative tree and then we define its one step transition probabilities. Next, given a particular decision rule, we determine the long run fraction of time that the tree under study is unhealthy. Finally, we ascertain the long run fraction of mandatory replacements that occur when the tree under consideration is dead.

1. Introduction

All over the world, distinct varieties of fruits and nuts, such as oranges, mangoes, peaches, and pecans are grown on orchards. As such, there is no denying the fact that orchards are now significant generators of agricultural revenue for their owners and/or managers in many different parts of the world. The work of Perrings (1987, 1996), Batabyal (1999), and Batabyal and Yoo (2007a, 2007b) tells us that the present and the future behavior of an orchard is governed by factors that are partly economic, and partly ecological, in nature. The economic factors are not difficult to comprehend. Specifically, orchards are the result of human actions that have been taken to encourage the commercial and the non-commercial consumption of fruits and nuts. The ecological factors affecting orchards include droughts, inclement weather, and, most notably for our purpose, invasive species. Using the parlance of modern researchers, orchards are *jointly determined* ecological-economic systems.

The current literature on orchards has studied many aspects of their management. For instance, Ilbery and Bowler (1995) have discussed the factors that have resulted in the selective adoption of the "Horticultural Buildings and Orchard Replanting Scheme" by farmers and orchard growers in England. Davis *et al.* (1997) have shown that the unevenness in yields that is routinely encountered in peach production in the United States can be decreased by increasing the distances between peach orchards. Park and Florkowski (2003) have examined the effect that orchard conditions and other factors have on the decision to utilize new varieties in peach production in the United States. Darroch *et al.* (2004) have analyzed the impact of alternate investment decisions on the profitability of apple orchard owners in South Africa. Florkowski and Sarmiento (2005) have analyzed the connection between the prices received by pecan growers and the characteristics of their orchards in the United States. Mouron *et al.* (2006) contend that, if the goal of sustainable farming in Switzerland is to be achieved, then cognitive competencies, such as distributional, conditional, and non-linear thinking, are crucial when knowledge from tree and orchard management is integrated at the farm level. Batabyal and Yoo (2007a) contend that judicious orchard management involves solving an optimization problem in which an economic objective function is minimized subject to an ecological constraint which says that the probability that the orchard under study is not Holling resilient[86] is at most as large as an exogenously given value. Finally, in a paper that is similar in spirit to our chapter, Batabyal and Nijkamp (2008a) have used a renewal theoretic approach to show that the critical number of trees in an orchard that must be infected and diseased before the manager takes his prophylactic tree removal and replanting actions is the solution to a particular minimization problem.

The papers we have just discussed in the previous paragraph have certainly enhanced our understanding of many aspects of orchard

86See Batabyal and Yoo (2007a) and the citations in this paper for additional details on the concept of resilience in the sense of C.S. Holling.

management. Even so, it is important to emphasize the following two points. First, the growth of and the output from orchards are both dynamic and stochastic in nature. This tells us that sustainable orchard management has everything to do with the *management of uncertainty* over time. Second, given the work of Acquaye *et al.* (2005) and Pierre *et al.* (2006), it is now clear that orchards are periodically *attacked* by one or more *invasive* species.[87] Therefore, efficient orchard management ought to clearly account for this aspect of the management problem. Now, even though the Batabyal and Yoo (2007a) and the Batabyal and Nijkamp (2008a) papers mentioned in the preceding paragraph have begun the process of modeling the above stated two points explicitly, the systematic study of these two salient points in the context of orchard management is still in its infancy.

Given this state of affairs, we focus on a representative tree in a stylized orchard and show how the theory of discrete time Markov chains[88] can be used to construct a model of this tree that is to be managed in the possible presence of one or more noxious invasive species. A discrete time Markov chain is a stochastic process that is the

87 Citrus canker and the Mediterranean fruit fly are two examples of invasive species that periodically have a deleterious effect on fruit orchards in Florida, in the United States. See Acquaye *et al.* (2005) and Pierre *et al.* (2006) for additional details on this point. In addition, invasive species represent one kind of potential harm that orchard managers have to episodically deal with. Other kinds of possible damage include droughts, inclement weather, and high winds.

88 See Ross (2003, pp. 181-268) and Tijms (2003, pp. 81-139) for textbook treatments of discrete time Markov chains.

simplest possible generalization of a sequence of independent random variables. In addition, a discrete time Markov chain is a random sequence in which the dependency of the successive events goes back only one time unit. Put differently, the amount of time a discrete time Markov chain spends in a particular state before making a transition to some other state is unity.

We carry out three orchard management related tasks. First, we specify an appropriate discrete time Markov chain theoretic model for our representative tree and then we define its one step transition probabilities. Next, given a particular decision rule (on which more in section 2.2 below), we determine the long run fraction of time that the tree under study is unhealthy. Finally, we ascertain the long run fraction of mandatory replacements that occur when the tree under consideration is dead.[89]

The rest of this chapter is organized as follows. First, section 2.1 provides a brief description of discrete time Markov chains. Second, section 2.2 delineates the stylized orchard, a representative tree in this orchard, and the decision rule to be used by the orchard manager to determine when to remove and replace trees. Third, section 2.3 specifies

89The reader should consult Batabyal and Nijkamp (2008b) and Rohlin and Batabyal (2005) for examples of probabilistic modeling in other economic settings.

a particular discrete time Markov chain model for the representative orchard tree and then it stipulates the pertinent one step transition probabilities. Fourth, section 2.4 computes the long run proportion of time that the representative tree under consideration is unhealthy. Fifth, section 2.5 calculates the long run proportion of mandatory tree replacements that occur when the representative tree under study is dead. Finally, section 3 concludes and discusses potential extensions of the research described in this chapter.

2. The Theoretical Framework

2.1. Preliminaries

Consider a stochastic process $\{X_n, n=0,1,2,\ldots\}$ that takes on a finite or countable number of possible values. Following Ross (2003, pp. 181-185), we shall denote this set of possible values by the set of non-negative integers $\{0,1,2,\ldots\}$. If $X_n=i$, then we shall say that the stochastic process is in state i at time n. We assume that when this stochastic process is in state i, there is a constant probability p_{ij} that it will next be in state j. Mathematically, this means that

$$p\{X_{n+1}=j / X_n=i, X_{n-1}=i_{n-1},\ldots, X_1=i_1, X_0=i_0\}=p_{ij} \quad (1)$$

for all states i_0, $i_1,\ldots, i_{n-1}$, i, j and all $n \geq 0$. A stochastic process that satisfies equation (1) is known as a discrete time Markov chain. In words, equation (1) is telling us that for a Markov chain, the conditional probability distribution of any future state X_{n+1}, given the past states $X_0, X_1,\ldots, X_{n-1}$ and the present state X_n, is *independent* of the past states and depends *only* on the present state.

Given that the stochastic process (discrete time Markov chain) is in state *i,* the constant p_{ij} denotes the probability that this stochastic process will next make a transition into state j in one step. Note that, because probabilities are non-negative, and because the stochastic process under consideration must make a transition into some state, we have

$$p_{ij} \geq 0\,,\ i, j \geq 0 \text{ and } \sum_{j=0}^{\infty} p_{ij} = 1\,,\ i = 0, 1, 2,\ldots \qquad (2)$$

With this description in place, let us now move to the specifics of the orchard management problem that is the centerpiece of this chapter.

2.2. The orchard and a representative tree

Consider a fruit orchard such as an apple or a pear orchard that consists of a finite number of practically identical trees. Consistent with our observation in footnote 87 in section 1, the trees in this orchard are vulnerable to attack by one or more invasive species, such as the Citrus canker or the Mediterranean fruit fly. As a result, at the beginning of each time period, the orchard manager inspects the various trees in his orchard to determine their health status. In what follows, we shall study the orchard management problem in a stochastic environment by focusing on a representative tree in this orchard. The reader should understand that because all the trees in this orchard are practically identical, the full orchard management problem is, in effect, a scaled up version of this individual tree management problem.

Now, focusing on our representative tree, an invasive species may attack or infect this tree at any point in time. Therefore, at the beginning of each time period (such as a day), the particular tree that our orchard manager has chosen to focus on is inspected to determine its health status. As a result of this inspection, the tree in question is deemed to be in a particular health state. Let us describe the N possible health states with the letter i where i=1,2,…,N. In this description, health state i is better than health state i+1. In particular, health state i=1 is the healthiest possible state and health state i=N is the state in which the tree under study is dead, i.e., it is the worst possible health state.

If the tree under study is dead, i.e, if it is in health state i=N, then this tree must be removed and replaced with a new tree. We suppose that this *mandatory* removal and replacement activity takes two time periods. If the tree is in some intermediate health state $i<N,$ then our orchard manager faces a choice between *preventively* removing and replacing this tree or letting this tree grow for the present time period. We suppose that preventive tree removal and replacement takes one time period. A new or replacement tree is uninfected and in the best possible health state. In symbols, this new tree is in health state i=1.

Suppose that our tree's current health state is i. Then, as a result of possible attack by one or more invasive species, in the next time period, its health state will be j with probability q_{ij} where q_{ij}=0 for $j<i$.[90] We suppose that our orchard manager uses the following decision rule to determine when to preventively remove and replace the tree under consideration. This tree is removed and replaced with a new tree only when the health state of the tree is greater than a critical value m, where m is a positive integer with the property that $1\leq m\leq N$. This completes our description of the stylized orchard and a representative tree in this orchard. We now specify a particular discrete time Markov chain for the representative orchard tree under study and then we write down the relevant one step transition probabilities.

90The reader will note that in our modeling of the underlying problem, the health of the representative orchard tree changes in a probabilistic manner, but the direction of this change is negative. Put differently, the health of this tree declines stochastically. In this way of modeling the underlying problem, we are constricting the state-space of the pertinent discrete time Markov chain. This issue and alternate formulations are discussed in section 3.

2.3. The Markov chain theoretic model

Recall from the section 2.2 discussion that the orchard manager's tree removal and replacement activity takes *two* time periods when the representative tree under study is dead, i.e., it is in health state $i=N$. Therefore, to model our representative tree with a discrete time Markov chain, we will first need to define an auxiliary state. Let this auxiliary state be 0. Further, let X_n denote the health state of the representative tree at the beginning of the *nth* time period. Then, we shall say that the stochastic process (tree) is in state *i,* with $1 \leq i \leq N$, when its *health state* is *i* and no mandatory removal and replacement is in progress. In addition, we shall say that the stochastic process (tree) is in the auxiliary state 0 when a mandatory tree removal and replacement has been in progress for one time period. We can now say that the stochastic process $\{X_n\}$ denoting our representative tree is a discrete time Markov chain.

We proceed to the one step transition probabilities in three parts. First, let us consider the intermediate health states that are at most as good as the critical health state *m.* These are the health states for which, in symbols, we have $1 \leq i \leq m$ and $j \geq i$. For these states, the pertinent transition probabilities are

$$p_{ij} = q_{ij}, \; 1 \leq i \leq m, \; j \geq i \tag{3}$$

Next, let us consider the health states that are no better than the critical health state m. Symbolically, these are the states for which we have $m<i \leq N\text{-}1$. Given the managerial decision rule (see the previous section 2.2), the one step transition probabilities are

$$p_{i1}=1,\ m<i\leq N\text{-}1. \tag{4}$$

Finally, we consider the transitions from health states N to 0 and from 0 to 1. For these two cases, the one step transition probabilities clearly are

$$p_{N0}=p_{01}=1. \tag{5}$$

For the discrete time Markov chain $\{X_n\}$ that we are studying, the reader will note that all the remaining transition probabilities not described by equations (3), (4), and (5), are zero. Our next task is to compute the long run fraction of time for which the representative tree under consideration is unhealthy.

2.4. Preventive tree removal and replacement

To determine this long run fraction, we will need to work with the so called limiting or equilibrium probabilities for our discrete time Markov chain.[91] These limiting probabilities give us the long run

91 See Ross (2003, pp. 200-213) and Tijms (2003, pp. 96-106) for textbook accounts of these probabilities.

proportion of time that our discrete time Markov chain is in a particular state. Now, for any state i, where i=0,1,…,N, let π_i be the limiting probability of being in that state. Then, these limiting probabilities can be computed by solving the so called balance equations.[92] In our case, the three sets of balance equations are

$$\pi_0 = \pi_N, \ \pi_1 = \pi_0 + \pi_1 q_{11} + \sum_{i=m+1}^{N=1} {}_i, \text{ and } \pi_j = \sum_{i=1}^{\min(j,m)} {}_i q_{ij}, \ 2 \leq j \leq N, \qquad (6)$$

along with the normalizing equation

$$\sum_{j=0}^{N} \pi_j = 1. \qquad (7)$$

With the computation of these limiting probabilities for our discrete time Markov chain $\{X_n\}$ out of the way, we are now in a position to state the long run fraction of time for which the representative tree under consideration is unhealthy. The limiting probabilities given above and some thought together tell us that the fraction we seek is given by

92See Ross (2003, pp. 370-372) and Tijms (2003, pp. 100-102) for textbook treatments of balance equations and their uses.

$$\textit{Long run fraction of time tree is unhealthy} = \pi_0 + \sum_{j=m+1}^{N} \pi_j. \tag{8}$$

The implications of equation (8) for orchard management are straightforward. First, we see that the manager can *reduce* the long run fraction of time our tree is unhealthy by "weakening" the decision rule he is using to determine when the tree under study ought to be preventively removed and replaced. From a practical standpoint, this "weakening" will involve *raising* the value of m so that this critical health state is now closer to the worst health state $i=N$. Second, note that as m approaches N, i.e., as our manager chooses to focus *only* on mandatory tree removal and replacement, the long run fraction of time our tree is unhealthy is, in a probabilistic sense, as small as it can possibly be. We now proceed to our last task in this chapter and this involves ascertaining the long run fraction of tree replacements that occur when the representative tree under consideration is dead.

2.5. Mandatory tree removal and replacement

To calculate this long run fraction, we shall once again make use of the limiting probabilities for our discrete time Markov chain $\{X_n\}$. Further, we will need to focus on the *average* number of mandatory tree replacements starting per unit time and on the *average* number of tree replacements starting per unit time. Now, using these two averages, some thought will convince the reader that the long run fraction of all tree removal and replacements that are mandatory is given by a particular ratio and that ratio is

$$\frac{average\ number\ of\ mandatory\ replacements\ starting\ per\ unit\ time}{average\ number\ of\ replacements\ starting\ per\ unit\ time}. \tag{9}$$

Our mathematical analysis thus far in this chapter tells us that the numerator of the ratio in (9) is given by π_N Similarly, the denominator of the ratio in (9) is given by $\sum_{j=m+1}^{N} \pi_j$. Now, combining these two expressions, we get

Long run fraction of replacements that are mandatory=

$$\frac{\pi_N}{\sum_{j=m+1}^{N} \pi_j}. \tag{10}$$

Recall from the discussion in section 2.2 that mandatory tree replacements require two time periods whereas preventive tree replacements require only one time period. From this we can reasonably suppose that, in general, mandatory replacements are more costly than preventive replacements. Given this situation, it stands to reason that, *ceteris paribus*, our orchard manager will want to *minimize* the ratio given in the right-hand-side (RHS) of equation (10). Now, contrary to our finding in section 2.4, this minimization can be undertaken by "strengthening" the decision rule being used by our manager to determine when he should preventively remove and replace the tree under study. From a practical perspective, this minimization will involve *lowering* the value of m so that this critical health state is now closer to the best health state i=1. Further, as m approaches 1, i.e., as our manager chooses to focus on preventive tree removal and replacement, the long run fraction of tree replacements that are mandatory will, in a probabilistic sense, be as small as it can possibly be. This completes our discussion of mandatory tree removal and replacement.

3. Conclusions

In this chapter, we focused on a representative tree in a stylized orchard and then we showed how the theory of discrete time Markov chains can be used effectively to construct a model of this tree that is to be managed in the possible presence of one or more noxious invasive species. Specifically, we carried out three orchard management related tasks. First, we specified a suitable discrete time Markov chain theoretic model for our representative tree and then we defined its one step transition probabilities. Next, given a particular decision rule, we determined the long run fraction of time that the tree under study is unhealthy. Finally, we ascertained the long run fraction of replacements that occur when the tree under consideration is dead.

The analysis in this chapter can be extended in a number of directions. Here are two suggestions for extending this chapter's research. First, using the results obtained in sections 2.4 and 2.5, it would be useful to set up and solve an optimization problem whose solution gives the orchard manager the optimal value of the critical health state Second, consistent with our observation in footnote 90, it would be instructive to analyze a model of orchard management in which the health of trees can decline and improve probabilistically. Studies of the optimal management of orchards that incorporate these features of the problem into the analysis will provide additional insights into a management function that has significant economic and ecological implications.

References

Acquaye, A.K.A., Alston, J.A., Lee, H., and Sumner, D.A. 2005. Economic consequences of invasive species policies in the presence of commodity programs: theory and application to citrus canker, *Review of Agricultural Economics*, 27, 498-504.

Batabyal, A.A. 1999. Aspects of the optimal management of cyclical ecological-economic systems, *Ecological Economics*, 30, 285-292.

Batabyal, A.A., and Nijkamp, P. 2008a. Optimal resource management in the presence of a deleterious alien species: A stochastic model for an orchard, *Letters in Spatial and Resource Sciences*, 1, 107-115.

Batabyal, A.A., and Nijkamp, P. 2008b. Is there a tradeoff between average patent pendency and examination errors? *International Review of Economics and Finance*, 17, 150-158.

Batabyal, A.A., and Yoo, S.J. 2007a. A probabilistic approach to optimal orchard management, *Ecological Economics*, 60, 483-486.

Batabyal, A.A., and Yoo, S.J. 2007b. A stochastic analysis of the Holling resilience of an orchard, *Ecological Economics*, 64, 1-4.

Darroch, M.A.G., Hardman, P.A., and Ortmann, G.F. 2004. An assessment of apple orchard investments in South Africa under uncertainty and irreversibility, *South African Journal of Economic and Management Sciences*, 7, 601-615.

Davis, S.B., Price, J.T., Wetzstein, M.F., and Rieger, M.W. 1997. Reducing yield variation in peach orchards by geographic scattering, *American Journal of Agricultural Economics*, 79, 1119-1126.

Florkowski, W.J., and Sarmiento, C. 2005. The examination of pecan price differences using spatial correlation estimation, *Applied Economics*, 37, 271-278.

Ilbery, B.W., and Bowler, I.R. 1995. Horticultural change and the horticultural buildings and orchard replanting grant scheme in England, *Environment and Planning C*, 13, 67-78.

Mouron, P., Scholz, R.W., Nemecek, T., and Weber, O. 2006. Life cycle management on Swiss fruit farms: Relating environmental and income indicators for apple growing, *Ecological Economics*, 58, 561-578.

Park, T.A., and Florkowski, W.J. 2003. Selection of peach varieties and the role of quality attributes, *Journal of Agricultural and Resource Economics*, 28, 138-151.

Perrings, C. 1987. *Economy and Environment.* Cambridge University Press, Cambridge, UK.

Perrings, C. 1996. Ecological resilience in the sustainability of economic development, in S. Faucheux, D. Pearce, and J. Proops, (Eds.), *Models of Sustainable Development.* Edward Elgar, Cheltenham, UK.

Pierre, R., Spreen, T.H., and Moss, C.B. 2006. Invasive species and biosecurity cost: Cost of monitoring and controlling Mediterranean fruit flies in Florida, *Journal of Agricultural and Applied Economics*, 38, 337-343.

Rohlin, S.M., and Batabyal, A.A. 2005. A theoretical perspective on managed rangelands and irreversible states, *International Review of Economics and Finance*, 14, 487-494.

Ross, S.M. 2003. *Introduction to Probability Models*, 8th edition. Academic Press, San Diego, CA.

Taylor, H.M., and Karlin, S. 1998. *An Introduction to Stochastic Modeling*, 3rd edition. Academic Press, San Diego, California.

Tijms, H.C. 2003. *A First Course in Stochastic Models.* Wiley, Chichester, UK.

www.ingramcontent.com/pod-product-compliance
Lightning Source LLC
LaVergne TN
LVHW020412220926
848484LV00016B/2553/J
9780988919310